Topics in Mining, Metallurgy and Materials Engineering

Series editor

Carlos P. Bergmann, Porto Alegre, Brazil

"Topics in Mining, Metallurgy and Materials Engineering" welcomes manuscripts in these three main focus areas: Extractive Metallurgy/Mineral Technology; Manufacturing Processes, and Materials Science and Technology. Manuscripts should present scientific solutions for technological problems. The three focus areas have a vertically lined multidisciplinarity, starting from mineral assets, their extraction and processing, their transformation into materials useful for the society, and their interaction with the environment.

More information about this series at http://www.springer.com/series/11054

Dionisio da Silva Biron · Venina dos Santos
Mara Zeni

Ceramic Membranes Applied in Separation Processes

Springer

Dionisio da Silva Biron
Universidade de Caxias do Sul
Caxias do Sul-RS, Rio Grande do Sul
Brazil

Mara Zeni
Universidade de Caxias do Sul
Caxias do Sul-RS, Rio Grande do Sul
Brazil

Venina dos Santos
Universidade de Caxias do Sul
Caxias do Sul-RS, Rio Grande do Sul
Brazil

ISSN 2364-3293 ISSN 2364-3307 (electronic)
Topics in Mining, Metallurgy and Materials Engineering
ISBN 978-3-319-58603-8 ISBN 978-3-319-58604-5 (eBook)
DOI 10.1007/978-3-319-58604-5

Library of Congress Control Number: 2017940233

Printed on acid-free paper

This Springer imprint is published by Springer Nature
The registered company is Springer International Publishing AG
The registered company address is: Gewerbestrasse 11, 6330 Cham, Switzerland

Preface

This book contains an overview of *Ceramic Membranes Applied in Separation Processes*. Membranes have been largely studied in recent years. They have great potential in several areas such as food, chemistry, pharmaceutical industries, and water and wastewater treatment. The separation processes by membranes are considered one of the most innovative technologies. These processes are considered fast and economical, occupy less space area, highly selective, flexible and compatible in most cases with other treatment processes. In general, the study of membranes can be divided into four major topics, materials for manufacture, study of membrane morphology, membrane modules, and the process. The study aims to achieve high performance, reducing costs with raw materials and process, defining its implementation and the reduction of environmental impacts. In selecting materials for the manufacture of membranes, ceramic materials (alumina, zirconia, titania, and silica) compared to polymeric materials have advantages such as longer life, high strength, and ease of cleaning. This book is aimed at students, researchers, and engineers who search for general knowledge in the field of membrane technology, particularly in the use of ceramic membranes.

The following chapters present several relevant aspects in the process of membrane separation, involving ceramic materials. Chapter 1 presents a brief introduction to the topic, Chap. 2 deals with an overview of some of the main processes of membrane separation, Chap. 3 presents the main raw materials used in the preparation of ceramic membranes as well as the main techniques of obtaining, Chap. 4 presents some techniques generally used in the characterization of ceramic membranes, Chap. 5 presents some applications of the ceramic membranes and, finally, Chap. 6 presents the ceramic membrane modules.

We wish that the following text guides the reader in a clear and objective way and that the information present will be appropriate for professionals and students who want to access knowledge focused on the area of ceramic membranes applied in separation processes.

Finally, we would thank the staff of Springer Verlag for their professional guidance in regard to this book.

Caxias do Sul-RS, Brazil
March 2017

Dionisio da Silva Biron
Venina dos Santos
Mara Zeni

Contents

Chapter 1
Introduction

Membrane separation processes using membranes have proved to be one of the most attractive and promising fields of technological research in recent decades. These processes consist of a technology that has been being developed and designed to perform physical or physical–chemical separations.

Although many of the processes involving membranes are liquid–fluid, there is also gas–liquid separation processes and gas–gas, although they are recently in development and their implementation is still not widespread. Among the various processes, hemodialysis is a milestone in the use of synthetic membranes, as half of all sales of membranes can be accounted for by this application. Still, the conquest of space membrane technology in different areas, such as in cases of water and effluent treatment, is based on its three advantages over conventional treatment techniques:

- the separation is achieved without the need for a phase change, and therefore it is more energy efficient than the distillation;
- little or no accumulation occurs in the process, therefore, it operates continuously under steady-state condition without the need of regeneration cycles, unlike adsorption separation processes;
- adding little or no chemicals is required, unlike conventional clarification that generally relies on the addition of chemicals (coagulants and flocculants).

Membrane separation processes have advantages when compared to conventional processes: energy saving, greater selectivity, separation of thermolabile compounds, and simplicity of operation and escalation. However, the main concerns involving membrane separation processes are the performance loss due to the filling of the membrane and the costs related to its cleaning and replacement. Still, the deposition of material on and within the structure of the membrane, called fouling, is not easily removed by pressure increase or backwashing. Consequently, there is a loss of permeability to the constant pressure or the need of increasing the pressure to keep the permeate flow constant and, thus, change the solute retention.

D. da Silva Biron et al., *Ceramic Membranes Applied in Separation Processes*,
Topics in Mining, Metallurgy and Materials Engineering,
DOI 10.1007/978-3-319-58604-5_1

In processes involving liquid effluents, the processes with membranes allow the removal from dissolved ions to the bacteria present in the water. The operating conditions, such as transmembrane pressure and the material to be withheld, will depend on the need imposed for each process.

One way to observe the scientific and technological importance of membrane separation processes is to check the number of publications made over the past few years. In 1998 were published around of 9,394 papers and, in 2016 this value increased to 50,226 papers, it equivalent to five time more. These studies related to membrane separation processes tend to increase over the years, because of the new scientific discoveries, especially with the nanotechnology arrival. The increase along this period is reflective to the need of searching for new technologies that have mainly the economic and ecologically correct bias (www.sciencedirect.com).

Currently, in the literature, an enormous range of research in membrane separation processes involving three major types of membranes can be observed:

- polymeric membranes
- ceramic membranes
- composite membranes

The first membranes made and commercialized were characteristic of polymeric material. The use of polymeric materials in the process of making such membranes occurred primarily due to their low cost and easy handling. However, due to need for materials with properties that allow the usage of membrane filters to more aggressive environments, such as high pressure, acidic and basic media, high temperatures, etc., the insertion of new materials to this technology was allowed, as ceramic and composite materials. Ceramic membranes, compared to polymeric membranes, have as characteristic to be more resistant to pressure, high temperatures, and to be resistant to a wide pH range (0–14), etc. On the other hand, they have high cost of manufacture, are limited to various settings, and usually are made in the form of tubes (mono and multi-channel). Regarding composite membranes, they usually have more than one material in their composition, allowing a union of different materials' properties, providing features inherent to the specific process which is being submitted. Composite membranes can provide reduction of manufacturing costs without reducing the degree of quality of the final product and even intensify the process properties, such as greater permeate flow and retention; besides presenting larger transmembrane pressure resistance, greater resistance to temperature and pH variations.

Polymeric membranes for microfiltration are obtained by various processes, including stretching, phase inversion (solvent casting), track etching, and sintering. In case of ultrafiltration membranes, the usual method for preparation is the phase inversion, as it is for reverse osmosis and nanofiltration. Figure 1.1 shows a micrograph of a PVDF polymer membrane with a pore size of about 0.5 μm.

Polymeric membranes can be applied in various areas, such as in the pharmaceutical industry, beverage industry, water and waste treatment, petroleum industry,

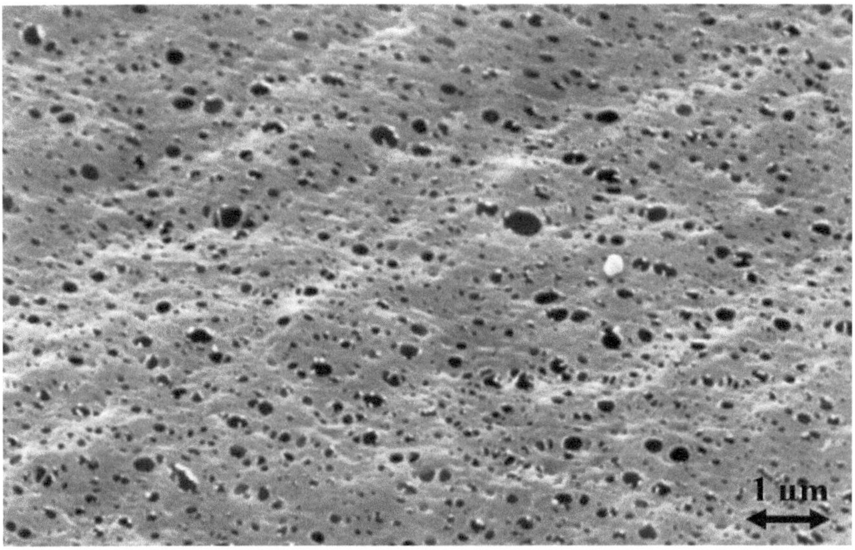

Fig. 1.1 Micrograph of a PVDF polymer membrane obtained by phase inversion process with pore size of about 0.5 μm

medicine, and other areas. Their wide application, as already mentioned, is due to their low cost and easy handling.

Figure 1.2 shows the number of publications made over the last few years about membrane separation processes involving ceramic membranes. In the year 1998, the number of publications was 1.065 and this number was exceeded in 2011, reaching an amount of 2.490 publications, which corresponds to 57% more published works. And in 2016, the number of publications came to 3.568, which corresponds to a percentage of 30% more in relation to the year of 2011 and 6.25% more compared to the previous year.

There are numerous methods of obtaining ceramic membranes. And this includes structural leaching or sublimation of one component of a mixture, slip casting, tape casting, extrusion, pressing, pyrolysis, sintering of suspensions that contain ceramic powders and various additives, sol–gel synthesis, hydrothermal treatment, anodic oxidation, and chemical vapor deposition. Figure 1.3 shows a micrograph of an alumina ceramic membrane obtained by the process of extrusion. The main materials used in the preparation of ceramic membranes include the α-alumina, mullite, zirconia, silica, and titania.

Ceramic membranes are applied in various areas, such as chemical industry, biotechnology, metal industry, food industry, in the treatment of liquid and gaseous effluents, and other areas. Their application, even if limited, in some areas, due to their higher cost, has been shown to be promising and has win space in many other industry sectors.

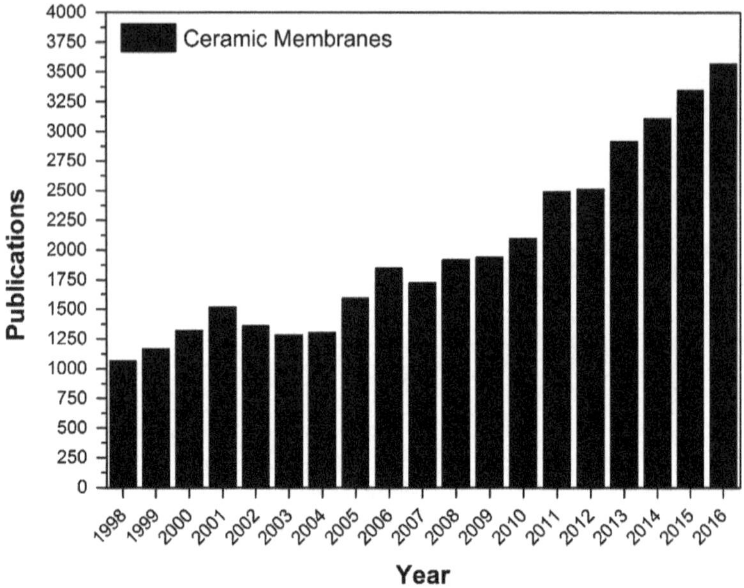

Fig. 1.2 Publications carried out on ceramic membranes from 1998 to 2016 (www.sciencedirect.com, accessed on October 21st, 2016)

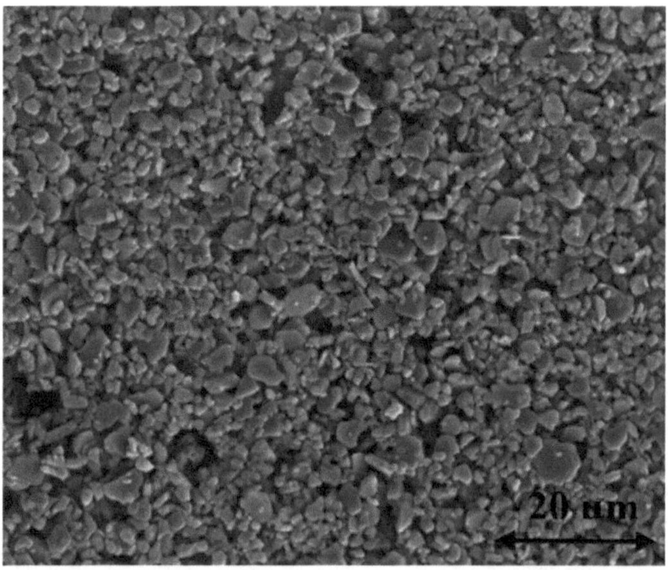

Fig. 1.3 Micrograph of an alumina ceramic membrane obtained by extrusion process with pore size of approximately 0.4 µm

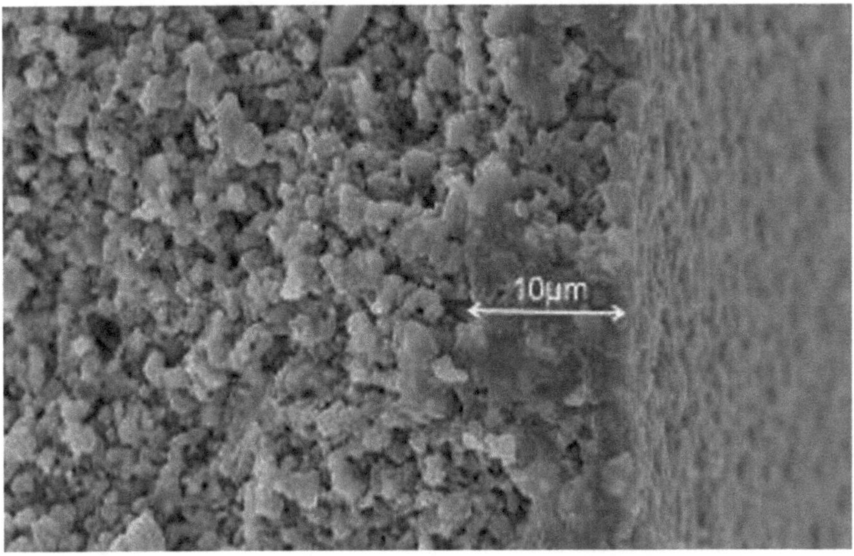

Fig. 1.4 Result of a micrograph of a composite membrane, where the deposition of polyamide 66 (PA66) was made, by the process of dip-coating, in α-alumina ceramic support

Composite membranes, as already mentioned, can be understood as a union of two or more materials with different characteristics that allow an improvement in process parameters. In general, composite membranes are prepared to increase selectivity and/or the permeability of the hybrid material. An example of composite membranes is the covering of a ceramic tube with a layer of polymer, using the process of dip-coating, with the objective of increasing the selectivity of the material. Figure 1.4 shows a micrograph of such example.

In this book, emphasis will be given on separation processes involving ceramic membranes. The next chapter regards an overview of the main processes of separation and where ceramic membranes are found. Then, in Chap. 3, which materials and preparation methods are used when preparing ceramic membranes are given. Chapter 4 presents the main techniques used in the characterization of these membranes, such as the porosimetry by mercury intrusion, rejection tests, permeability study, and others. Also, in Chap. 5, a few possible applications of ceramic membranes will be presented. Although polymeric membranes are more widely used on the market, we will see that ceramic membranes conquer more and more space in several applications. Last but not least, in Chap. 6 the modules used for these membranes will be presented, which are of vital importance in the construction of a plant in industrial scale.

Chapter 2
Overview Membranes Separations

The technology of membranes separation is a field that involves many processes. These processes are subdivided according to the driving force applied, the feeding phases, the permeate, and the pores' size. The applied driving force is understood as the necessary force for the feeding fluid to cross through the membrane until the permeate. The feeding phases, generally, are found in liquid or gaseous state. Another important aspect, which is considered crucial for the process, is the pores' size, as many separation processes involve the principle of size exclusion. Figure 2.1 shows the main processes of membrane separation which involve transmembrane pressure as driving force. According to the figure, we can see the pressure regions where microfiltration (MF), ultrafiltration (UF), nanofiltration (NF), and reverse osmose (RO) processes occur, as well as the size of the solute to be excluded for each process and material to be withheld.

In this chapter, the main processes in technology of membrane separation will be discussed succinctly. These processes are MF, UF, NF, and RO. Still, permeation of gas (PG) and the membrane distillation (MD) processes will be presented, in which ceramic membranes have been gained space.

2.1 Membrane Definition

To a better understanding of membranes separation processes, it is necessary to know what is a membrane. The complete and precise definition of membrane can be easily understood, as the synthetic material, which when used turns into a membrane, as well as the process of obtaining present a diversity of aspects to be considered. Generally, a membrane can be characterized as a barrier that separates or restrings partly or totally the passage of existing chemical species due to its characteristics. This membrane can be homogenous or heterogenous, had positive or negative charge, present neutral charge or even, be bipolar, symmetric or

D. da Silva Biron et al., *Ceramic Membranes Applied in Separation Processes*,
Topics in Mining, Metallurgy and Materials Engineering,
DOI 10.1007/978-3-319-58604-5_2

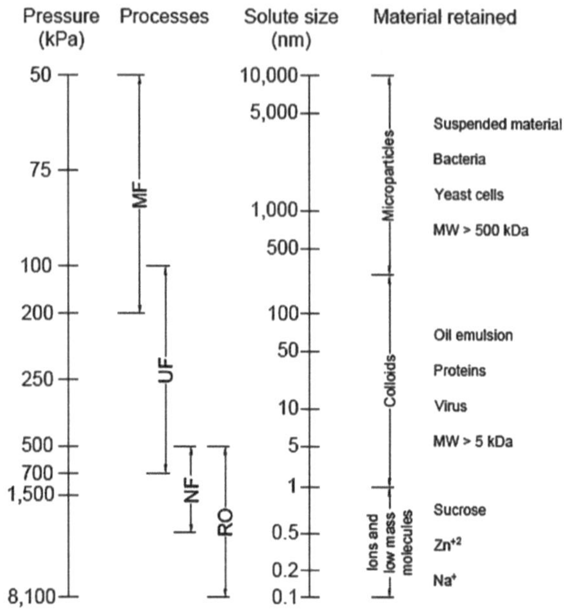

Fig. 2.1 Main membrane separation processes, involving transmembrane pressure as driving force (Singh 2006)

asymmetric, porous or dense and can be composed by various materials. Still, membranes can be polymeric, ceramic or composed. The material which it is done will depend on the necessities imposed and the kind of process to be used.

Figure 2.2 shows a presentation of a membrane and the possibilities which surround it, as, for example, the kind of material, the driving force, diverse existing configurations, the structure, and what is possible to separate.

The species took from the membrane can be called retentate or concentrate, while the species which cross the membrane structure are denominated permeate or filtered. The driving force will be used in membrane separation processes can be a gradient of applied pressure, gradient electric potential or concentration gradient.

2.2 Microfiltration

2.2.1 Microfiltration History

The beginning of microfiltration usage can be dated on the nineteenth century with synthesis of nitrocellulose, in 1845, by Schoenbein. On the following years, in 1855, Fick used collodium solutions (Ether-alcohol) to form the first nitrocellulose membrane. In 1906, the research in microfiltration membranes continued with Bechold, that obtained porous collodium membranes and measured the pores size by the bubble point test. In this same century, in around 1925, researchers like Bigelow, Gembering, Schoep, Brown, Zsigmondi, and Bachmann made significant

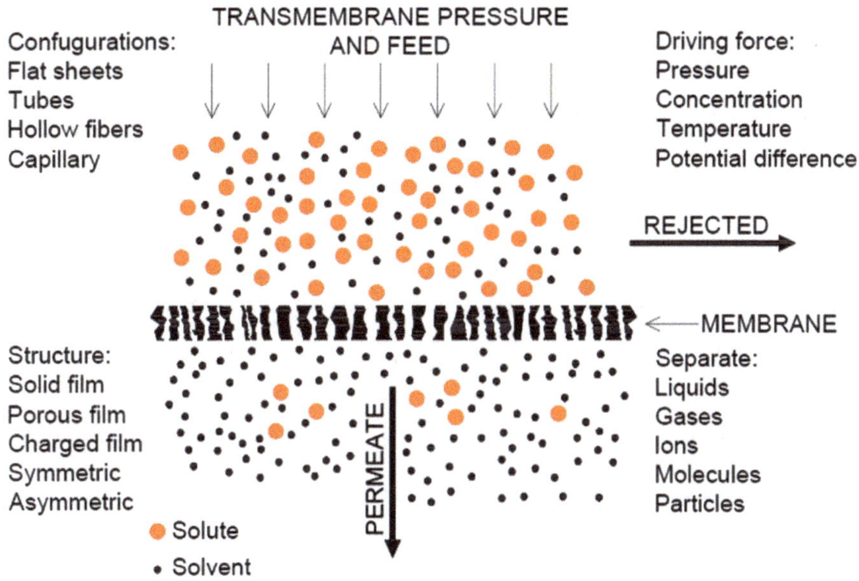

Fig. 2.2 Representation of the separation process involving membranes as selective barriers

advances on methodology of pores sizes' spreading and regulation. However, despites it high level of sophistication in controlling the size of the pores and microstructure, the authors still have no knowledge of the mechanisms in formation on membrane. The first big and great application of this microfiltration happened in the Second World War, where there was an urgent need for an efficient method as a detection of serious pathogenic bacteria.

After the great war (1947), US Joint Intelligence Objectives Agency sent Dr. Alexander Goetz to Germany to obtain information about the method of production of filter membranes. Goetz visited Membrane filter gesellschaft and based on his discoveries he was awarded with a contract by the US Chemical Corps to develop membranes. In 1950, Goetz had improved the method of membranes production, obtaining higher fluid taxes and more uniformed pore sizes. Based on developments from Goetz, the Lovell Chemical Company in Watertown, Massachusetts, adjudicated more contracts in 1952 to commercialize the production. In 1954, Lovell Chemical Company sold the manufacture to Millipore Corporation and other companies, which started to make membranes according to the German techniques.

2.2.2 *Microfiltration Process*

Microfiltration process can be understood as a technique of separation, able to remove particles with sizes on micrometers scale, as bacteria, yeast cells, colloids, and suspended particles. The fluid passage with the contaminants will occur through a membrane with pore diameter of 0.1–10 μm. This average of size covers a great variety of natural and industrial particles, as shown on Fig. 2.1. The separated particles in this process are much bigger than the solutes which are separated by reverse osmosis and ultrafiltration, consequently the osmotic pressure is paltry. The driving force in this process is the gradient of pressure formed between the two sides of membrane and can vary from 50 to 200 kPa (Ho and Sirkar 1992).

In this process, the membrane pore size is usually bigger than the processes of ultrafiltration or reverse osmose, consequently the microfiltration process has bigger permeated fluxes. During the microfiltration process, the imposed pressure, which is the driving force to the process, makes the fluid and the smaller diameter go through the membrane and be collected as permeated. The bigger particles are removed by the filtering and collected as concentrate. The mechanism which the particles are removed depends on the kind of filtering and of the kind of interaction with particles to be filtered (Cheryan 1998).

MF process can be done on two different ways: dead-end and crossflow. A classic example of dead-end flux is the use of MF cellulose nitrate or cellulose acetate membranes, diameter 0.45 μm, to the count of bacteria. In this configuration, the feed flux goes to the membrane surface in perpendicularly. As shown in Fig. 2.3, transmembrane pressure obligates the fluid passage through the membrane's pores, giving a clarified permeate. On dead-end process, the particles tend to accumulate along the filtration process, and, when a solute layer (cake layer) is formed, the permeate flux decreases drastically. As consequence, the filtration

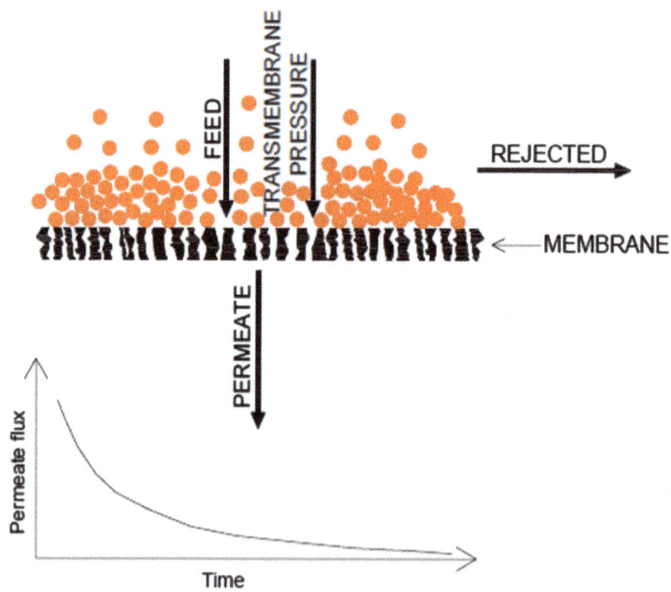

Fig. 2.3 Dead-end configuration

process shall be constantly interrupted so that the solute layer is removed or the membrane is replaced (Scott 1998).

Crossflow configuration appeared as an alternative to dead-end configuration. For the crossflow microfiltration the sieve mechanism is seen as dominant of the process. This configuration is often used on ultrafiltration, to the clarification of beverages in the food industry. Still, this configuration, differently from dead-end, is used for a feed with higher quantity of solute. The main advantage of this configuration, compared to dead-end, is the shear force provoked by the fluid which drains parallelly to a membrane surface, which provokes a drag and decrease of the cake layer. Consequently, relatively high permeate fluxes can be achieved and maintained for a long time. According to Fig. 2.4, the constant permeate flux is reached when the cake layer hits a determined thickness. However, in practice, the decrease on permeate flux is observed even when there is no accumulation of solute. Besides, a more commonly used module to the dead-end configuration is the plate and frame module, while, on crossflow configuration, it is the tubular module (Oyama and Stagg-Williams 2011; Meyer et al. 2015).

When the sieve mechanism on the microfiltration is dominant, a cake layer of rejected particles usually is formed on the membrane surface. The permeate flux by the pressure through the cake layer and the membrane can be described by Darcy's Law, Eq. 2.1.

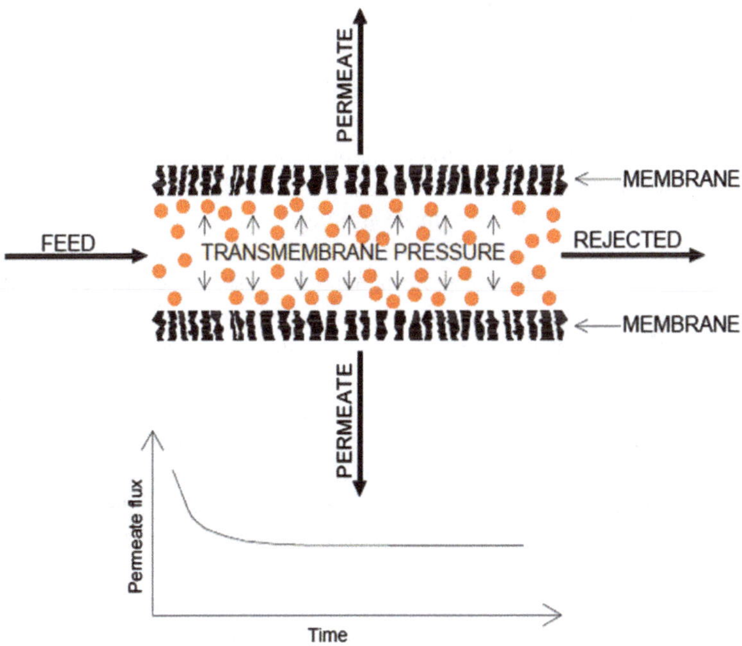

Fig. 2.4 Crossflow configuration

$$J = \frac{1}{A} \cdot \frac{dV}{dt} = \frac{\Delta P}{\mu \cdot (R_m + R_c)} \quad (2.1)$$

where J is the permeate flux, A is the membrane area, V is the permeate volume, t is the time of the filtration process, ΔP is the transmembrane pressure difference, μ is the absolute viscosity of the fluid, R_m the resistance given by the membrane, R_c is the resistance presented by the cake layer.

In case the membrane gets near a more porous environment formed by spheres which touch tangentially, the permeate flux can be calculated through Carman–Kozeny's model, Eq. 4.4.

Hydraulic permeability is another parameter to be calculated on microfiltration process. Hydraulic permeability can be understood as a measure of greater or lesser ease which the membrane offers to the passage of a determined solvent, and can be calculated by Eq. 2.2.

$$L_p = \frac{J}{\Delta P} \quad (2.2)$$

2.3 Ultrafiltration

2.3.1 Ultrafiltration History

The concretion of the beginning of ultrafiltration process coincides with the reverse osmose, around 1960. However, the word ultrafiltration emerged in 1856 with Schmidt, which carried out the filtration of proteins and Arabic gum with the usage of a membrane of animal origin. In 1906, Bechold produced collodium membranes with pores smaller than 0.01 μm, made from acetic acid and cellulose nitrate.

The discovery of an anisotropic reverse osmosis membrane, in 1959, opened doors to an anisotropic ultrafiltration membrane. Despite the differences between the membranes, their evolution process was given parallelly. Before that time, Zsigmondy produced one of the first patents in this area, about the preparation of flat collodium membranes. However, these membranes were only a curiosity of the academia, and were applied in the biotechnology area and in laboratory scale.

After the First World War, the American government got worried with water shortage before the end of the century. The United States Department of the Interior created the Office of Salina Water (OSW) and pawned substantial financial resources to the development of various separation processes to the desalination of water, which included processes with membranes. In this time, Loeb and Sourirajan developed asymmetric cellulose acetate membranes, which had smaller thickness, more selectivity and higher permeability rate, which reduced the process time and the permeate area. Besides, these membranes presented macroporous structure, responsible for its higher mechanical resistance.

2.3.2 Ultrafiltration Processes

The ultrafiltration process is a separation process based on the principle of size exclusion. The UF membrane pore size is in average 1–500 nm and is able to retain solutes which present molecular weight from 300 to 500,000 Da. The operating pressure is around 100–700 kPa. UF can reject proteins, such as trypsin, egg albumin and bovine serum albumin. One of the characteristics of UF membranes is your distinction in regards to your molecular weight cutoff (MWCO). Usually, the MWCO set the ability of membrane and rejection occurs when the membrane presents rejection equal or greater of 90%. Despite this present process deletion by particle size, often the membrane may have the ability of retaining smaller particles expected. This is mainly due to the formation of a layer of solute, despite contributing permeate flux decline also contributes in a resistance to passage of the solute. In fact, this characteristic hinders the capacity analysis of rejection of the membrane, so the trials involving the MWCO allow a better understanding of the real efficiency of rejection of the membrane. Other molecules involved in this process are the sugar, biomolecules, polymers, and colloidal particles (Ho and Sirkar 1992; Scott 1998).

The MWCO of any membrane can vary with the chemical characteristics of feed as well as with the factors such as molecular orientation and molecular configuration. However, the nature of the larger molecules that are normally separated by UF leads to significant practical differences between UF and RO processes. As a consequence of higher molecular weight of the species separated into a process of UF, osmotic pressure differentials are smaller. At the same time, the liquid phase diffusivity of these species is also lower. Therefore, the problems of fouling of membrane polarization and by concentration are more significant in UF (Scott 1998).

The permeate flow in UF process is given as Eq. 2.3. The difference of calculating the flow of permeated the process of MF for the UF is on the increase in osmotic pressure, which in the case of UF becomes significant.

$$Q = \frac{k \cdot (\Delta P - \Delta \pi) \cdot A}{l} \tag{2.3}$$

where Q is the volumetric rate of the solvent through the membrane, k is the coefficient of permeability of the membrane, ΔP is the difference of transmembrane pressure, $\Delta \pi$ is the osmotic pressure difference across the membrane, the membrane area and l is the thickness of the membrane. The permeate flow is equal to the ratio of the volumetric rate and membrane area (Q/A).

The main constraints permeated flow membrane separation processes are extensively studied, because they directly influence the process. These constraints are the polarization by concentration, fouling, and the resistance of the membrane.

The resistance of the membrane is caused by geometrical factors such as pore size and distribution, thickness, hydrophobicity, and membrane hydrophobicity. As on Eq. 2.4, you can calculate the resistance imposed by the membrane, whereas a new membrane and pure water as feed.

$$R_{\mathrm{m}} = \frac{\Delta P}{(\mu_{\mathrm{W}} \cdot J_{\mathrm{W}})} \tag{2.4}$$

where the index W refers to the properties of water.

Fouling resistance is the sum of the resistances of adsorption, the formation of layer cake and blocking pores. It can be calculated according to the Eq. 2.5, considering pure water flow measured after the filtering process.

$$R_{\mathrm{f}} = \frac{\Delta P}{(\mu_{\mathrm{W}} \cdot J_{\mathrm{W1}})} - R_{\mathrm{m}} \tag{2.5}$$

where W refers to the properties of water and J_{W1} concerns permeate flow measured after filtering.

Obtaining the resistance presented by polarizing layer by concentration can be obtained according to Eq. 2.6. To this end, it is necessary to determine the total resistance (RT) presented during the filtration, being the sum of the resistance of the membrane fouling and concentration polarization by layer. The total resistance is determined as Eq. 2.7.

$$R_{\mathrm{c}} = \mathrm{RT} - \mathrm{RF} - \mathrm{RM} \tag{2.6}$$

$$\mathrm{RT} = \frac{\Delta P}{\mu_{\mathrm{S}} \cdot J_{\mathrm{S}}} \tag{2.7}$$

where μ_{S} is the viscosity of liquid (in this case is the solution that contains the solution that will be filtered) and $\mathrm{J_S}$ is the permeate flow measured feed solution.

2.4 Nanofiltration

2.4.1 Nanofiltration History

The history of NF started around the late 70, when the RO membranes began to operate with a reasonable water flow, operating at relatively low pressures. So, with a reduction in operating pressure, which was traditionally high in RO, it was possible to establish a reduction in energy costs. In this way, membranes with smaller rejections of dissolved components, but with higher permeability to water, if presented with a big improvement for the technology of separation (Hilal et al. 2004). Despite several studies presented at this time, the use of the word nanofiltration, in commercial, was established by the Filmatec Corporation (in 1980), to define a region between ultrafiltration and reverse osmosis. The NF, also known as loose RO has pores with diameters of about 1 nm. The full definition of NF can be established by a number of features, which are quoted below (Paul and Jons 2016):

- pore diameters of less than 2 nm;
- passage of a significant amount of monovalent ions;
- substantially larger divalent ion rejection than monovalent;
- the molecular weight cutoff (MWCO) for neutral species is in the range of 150 and 2000;
- the rejection of neutral and positive ions refers mainly to size and shape.

In fact, the membranes with selectivity in the regions between UF and RO already existed in the years 60, however were termed as open, loose, low-pressure RO, intermediate RO/UF, or tight UF membranes. However, only from the years 80 the word nanofiltration became used firmly and explicitly used by several authors (Wang et al. 2011). Thus, different RO membranes, which have non-porous structure and a transport mechanisms for diffusion, NF membranes operate on porous and non-porous membranes interface, prevailing the mechanisms of diffusion and deletion by particle size.

2.4.2 Nanofiltration Process

In NF, as well as other membrane separation technologies, mass transit can lead to binding, accumulation or absorption of materials on the surfaces of the membrane and/or within the porous structure, causing a decline on the permeated flow throughout the period. The consequence of that, is the decrease of permeability and the need to increase the transmembrane pressure to maintain the desired flow, in addition to increased energy consumption, operating costs and increase the frequency of cleaning. Cleaning the membrane is the main form of mitigation of polarization layer formation by merger, the formation of gel layer and the effects of the phenomenon fouling, but a loss will occur in the lifetime of the membrane (Mohammad et al. 2015).

NF membranes exhibit an active layer, where the rejection occurs, which features pores with nanometer geometry, and can be characterized by different techniques. The main techniques used are those of gas adsorption and desorption, atomic force microscopy, transmission electron microscopy (TEM), and rejection of solutes. The hydrophilic/hydrophobic properties are constantly evaluated in this type of membrane for presenting important relevance of the membrane interaction with the solvent, which can be determined with relatively inexpensive techniques (Mohammad et al. 2015).

Another important feature presented by the membranes are the loading properties, which are the key to the understanding of the process and this variable depends on the nature, concentration, and pH of the solution. These electrical characteristics allow an increase in mass transport through the membrane and are significant in rejection of the solute (Wang et al. 2011). The NF membrane in contact with the aqueous solution are slightly charged due to the dissociation of functional groups or surface adsorption of solute. For example, NF membranes contain ionizable groups polymer such as carboxylic groups and sulfonic acid groups resulting in a charged surface in the presence of a feeding solution. Similar to the RO membranes, NF membranes are potent in the separation of inorganic salts and small organic molecules (Mohammad et al. 2015).

As well as the fouling, another drawback of NF membrane is the problem of controlling the reproducibility of the size and distribution of pores in the membrane (Wang et al. 2011).

The Nernst–Planck equation, Eq. 2.8, is extensively used to measure the transport of ionic species through the membrane pores of NF. The application of this equation was originally proposed by Schlogl for description of electrolyte transport in RO through ion exchange membranes. The equation is particularly useful for NF, once consideration is given to the transport mechanisms and parameters required, based on the properties of a real membrane. When considering the movement of unidirectional composites by membrane, and assuming ideal solutions, the transport equation for the species i can be written as (Chaabane et al. 2007):

$$j_{\mathrm{i}} = -\frac{C_{\mathrm{i}}.K_{\mathrm{i,d}}.D_{\mathrm{i},\infty}}{R \cdot T} \cdot \frac{\mathrm{d}\mu}{\mathrm{d}x} + K_{\mathrm{i,c}} \cdot C_{\mathrm{i}} \cdot V \tag{2.8}$$

where j_{i} is the ionic flow, C is the concentration, V is the velocity of the solvent and $K_{\mathrm{i,d}}$ and $K_{\mathrm{i,c}}$ are the obstacle factors to explain convection and diffusion within a confined space.

There are several variations of the equation that models the NF membrane transport. However, the proposed models are based on the Nernst-Planck equation and are variations based on a methodological solution, implications or extensions as the particular parameters (Mohammad et al. 2015).

Despite being the main polymer materials used in the preparation of NF membrane, some authors have reported the use of ceramic membranes for this purpose. Ceramic membranes of NF mainly use materials of titania, zirconia, silica-zirconia, hafnia, and alumina. Most of these nanofiltration membranes were prepared for the separation of non-aqueous solvents using sol-gel process, in which a mesoporous ceramic support is coated with a layer of a metallic oxide that determines the final size of the pores in the active layer. This procedure provides a great advantage in the pore diameter control through the appropriate choice of colloidal solutions at the stage of final coat. The molecular weight cut-off presented by these membranes is located between 200 and 1000 Da (Gitis and Rothenberg 2016).

2.5 Reverse Osmosis

2.5.1 Reverse Osmosis History

The first observations of the osmosis process were held in the eighteenth century, with studies of Nollet and Dutrochet. While these authors contributed to the first steps of the methodological observations involving membrane permeation, Fick and Graham, 1855 and 1866, respectively, were permeant species could be differentiated rates in transport across the membrane. Later, the osmotic pressure measurements were performed by Traube in 1867 and Pfeffer in 1877 and, in 1887, Van't Hoff used these concepts to explain the behavior of ideal solutions diluted. Reverse osmosis is a well-established technology of water purification, using semi-permeable membranes in the removal of ions and molecules of low molecular mass. The largest application of this process is for the desalination of water for consumption (Wang et al. 2011).

For decades, many authors have reported several studies involving the processes of osmosis and reverse osmosis. In 1959, in the work of Reid and Breton, with cellulose acetate films, it was possible to achieve values of 99% salt rejection of the water, however the permeate flows obtained were considered impractical at the time, reaching a maximum of 1.17 L/m^2 (Reid and Breton 1959). Still, these authors verified the influence of thickness of selective retention efficiency skin and permeate flow. In the following years, in 1962, Loeb and Sourirajan reported on their studies a high salt rejection with streams of water permeated movies using relatively high cellulose acetate. These studies, among others, formed the basis for the understanding and improvement of the process of reverse osmosis in later years.

2.5.2 Reverse Osmosis Process

When two solutions with different concentrations of solute are separated by a semipermeable membrane, is established a chemical potential difference. Of course, the phenomenon of osmosis is the diffusion of water through a semi-permeable skin to a region of lower concentration (greater chemical potential) to a region of higher concentration (lower chemical potential), until the chemical equilibrium is reached. To achieve the balance the difference in pressure between the two sides of the membrane is equal to the difference in osmotic pressure. To perform the reverse process it is necessary to apply a pressure difference greater than the osmotic pressure, which allows the separation of solute solvent. This phenomenon is called Hyperfiltration or reverse osmosis, Fig. 2.5 (Metcalf and Eddy 2014).

The following equations are used to describe the permeate flow submitted by RO membranes, as well as the calculation of the area of membrane and the number of stages required for a given application (Metcalf and Eddy 2014).

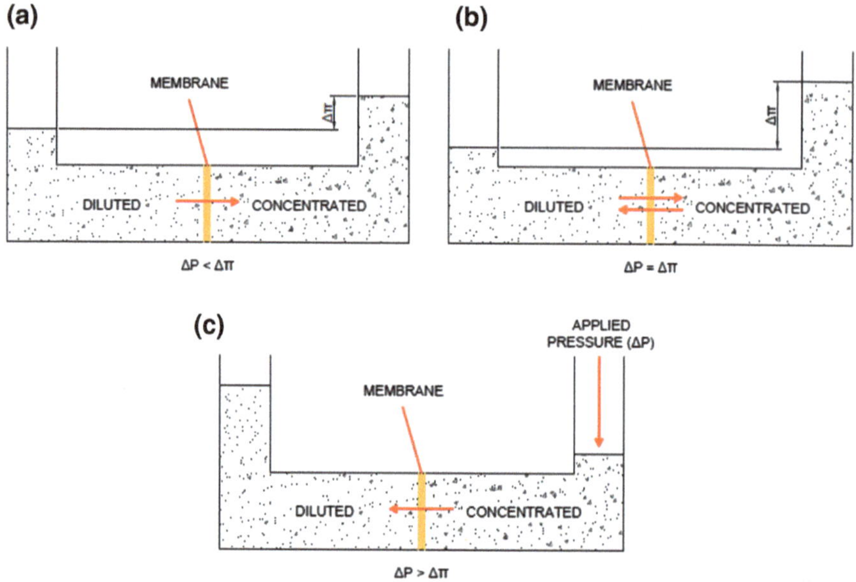

Fig. 2.5 Reverse osmosis process schema. **a** Natural process of osmosis, osmotic balance; **b**, **c** reverse osmosis

$$J = k \cdot (\Delta P - \Delta\pi) = \frac{Q_P}{A} \tag{2.9}$$

$$\Delta P = \left(\frac{P_f - P_c}{2}\right) - P_p \tag{2.10}$$

$$\Delta\pi = \left(\frac{\pi_f - \pi_c}{2}\right) - \pi_p \tag{2.11}$$

where J is the permeate flow ($m^3\ m^{-2}\ h^{-1}$), k is the mass transfer coefficient for the stream of water (depending on the water temperature, the characteristics of the membrane and the solute) ($m^3\ m^{-2}\ h^{-1}\ bar^{-1}$), ΔP is the difference of applied pressure (bar), P_f is the feed pressure (bar), P_c is the concentrated pressure (bar), P_P is permeated pressure (bar), $\Delta\pi$ is the osmotic pressure (bar), πf is the osmotic pressure of feeding (bar), π_c is osmotic pressure of concentrated (bar), π_p is the osmotic pressure of the permeate (bar), Q_p is permeated flow ($m^3\ h^{-1}$), and A is the area of the membrane (m^2).

In most cases, the solutes will pass through the membrane. This way, the solute flow can be calculated by Eq. 2.12 (Metcalf and Eddy 2014).

$$J_S = k_S \cdot \Delta C_S = \frac{(Q_P)(10^{-3}\ \mathrm{m}^3\ L^{-1}) \cdot C_P}{A} \tag{2.12}$$

$$\Delta C_S = \left(\frac{C_f - C_c}{2}\right) - C_p \tag{2.13}$$

where J_S is the solute mass flux (g m^{-2} h^{-1}), k_S is the mass transfer coefficient for the solute (m h^{-1}), Q_P is permeate flow (m^3 h^{-1}), A is the area of the membrane (m^2), C_P is the concentration of solute in permeated, ΔC_S is the variation of concentration of solute through the membrane (g m^{-3}), C_f is the concentration of the solute in the feed (g m^{-3}), and C_c is the concentration of solute in concentrated (g m^{-3}).

The osmotic pressure, for sufficiently dilute solutions can be calculated by Eq. 2.14. This equation was proposed by van't Hoff, a time to observe the osmotic pressure found that this had a similar behavior to an ideal gas. However, as the ideal gas equation to account for factors fixes for real gases, the van't Hoff equation allows, by van't Hoff factor, the fix for real solutions, Eq. 2.15.

$$\Delta\pi = n \cdot R \cdot T \cdot i \tag{2.14}$$

$$i = 1 + \alpha \cdot (q - 1) \tag{2.15}$$

where $\Delta\pi$ is the osmotic pressure (Pa), n is the solute concentration (mol m^{-3}), R is the ideal gas constant (8.314 Pa m^3 mol^{-1} K^{-1}), T is the absolute temperature of the solution (K), i is the correction factor of Van't Hoff, α is the degree of ionization, and q is the total number of ions released in the ionization of a compound.

Reverse osmosis is classified into three categories: The first, named RO, is for processes that involve high pressures, 5.6–10.5 MPa. The second involves low pressure processes of 1.4–4.2 MPa, and is called reverse osmosis. The osmosis of low pressure is applied on the desalination of brackish water. Finally, the third category is called nanofiltration or loose RO and operates at pressures from 0.3 to 1.4 MPa (Ho and Sirkar 1992). The nanofiltration was discussed in the previous chapter.

Currently, RO membranes are made of polymeric materials, such as cellulose acetate and polyamide. The most important are the membranes of acetate cellulose with homogeneous asymmetric morphology, the polyamide obtained by phase reversal and the cross-linked thin film composite (TFC) polyamides. Both the cellulose acetate membranes and polyamide possess a combination of economically viable high rejection and water flux. TFC membranes have higher added value than cellulose acetate membranes, but have higher flow and, therefore, operate at a lower pressure (Singh 2006; Lee et al. 2011).

The surface of the polymeric membranes of RO is considered non-porous, consisting of a polymeric network in which the solvent can be dissolved. RO membrane has a dynamic structure that allows the entry of water molecules. In fact, the dense skin has pores that are considered transitional channels. The main problems associated with polymeric membranes of RO is the excess fouling due to poor feed flow hydrodynamics, low resistance to oxidizing agents (such as chlorine), extensive use of pretreatment and chemicals associated with the generation of waste and lack of surface charge to reduce the fouling (Li 2007; Wang et al. 2011). Another problem associated with RO membranes is the decrease in the flow of permeated the solvent, a result of accumulation of solute molecules on the surface of the membrane, contributing to the increase in osmotic pressure and, as a result, the increase in energy costs.

Tubular and plate modules were the first modules used in the processes of RO. However, due to the low density packaging, these were replaced by modules that allow a higher ratio between the area of the membrane and the volume of the module. Currently, the most widely used are the hollow fiber and spiral wound modules, because they have the highest density packaging, which allows a considerable gain in the permeate flow. However, these modules feature ease of clogging and difficulty of cleaning.

Despite several advantages presented by ceramic membranes compared to the polymerics membranes, as resistance to oxidizing agents and solvents, high pressures resistance and among others; the costs involved in obtaining, low ability to retain ions and molecules of low molecular mass and low density of packaging make unfeasible the use of these membranes in the RO process.

On Li and collaborators' paperwork, the authors conducted a study using MFI-type zeolite membranes supported with α-alumina with the purpose of purification of water by the process of RO (Li et al. 2004). In this study, the authors found that the ion rejection and the flow of water depend on the charge density of the ions, the size and dynamic of hydrated ions diffusivity. Yet, the authors obtained as a result, for a 0.1 M solution of sodium chloride, with flow of 0.112 kg m^{-2} h^{-1}, a rejection of Na^+ of 76.7%, with transmembrane pressure applied to 2.07 MPa. In addition to NaCl, the authors evaluated the rejection of NH_4Cl, KCl, $CaCl_2$, and $MgCl_2$, getting rejections between 58.1 and 88.4%. Zeolite membranes have been mainly applied in gas separation and liquid pervaporation processes, but recently, dynamic simulation studies have shown that zeolite membranes are theoretically applicable in RO processes and can reach salt rejections of up to 100% (Lee et al. 2011).

2.6 Gas Separation

2.6.1 Gas Separation History

The major milestones of the process of separation of gases are, in large part, on its findings, mainly, of the characteristics and behavior of gases when applied to different pressures and temperatures.

The history of separation of gases can be dated from the findings of a French physicist called Guillaume Amontons in Thermometry and mathematics, which led to his assumption of absolute zero. In 1720, Gabriel Daniel Fahrenheit developed the idea of a temperature scale, which took your name. The Swedish Anders Celsius in 1741, formulated the temperature in degrees centigrade, where zero represents the freezing of water and 100 °C the temperature at which the water boiled. In this same century, the scholar Joseph Priestley managed to isolate a gaseous compound unknown, which Lavoisier later found to be oxygen, one of the ingredients of the Earth's atmosphere. Priestley also found ammonia, sulfur dioxide, nitrogen gas and later identified as carbon monoxide (Yampolskii and Freeman 2010).

In the nineteenth century, it was possible to liquefy industrial gases and check their behavior and characteristics. In 1823, Michael Faraday, would liquefy chlorine and studied the liquefaction and the characteristics of the ammonia. In the nineteenth century, it was possible to liquefy industrial gases and check your behavior and characteristics. In 1823, Michael Faraday, would liquefy chlorine and studied the liquefaction and the characteristics of the ammonia. In 1824, Sadi Carnot presented in his thesis findings which revealed that a thermal engine was more efficient, operating at your ideal, when two tanks (one hot and one cold) were used. In 1845, Thomas Andrews managed to establish the first isotherms, evaluating the behavior of the carbon gas at different temperatures and pressures. He found that the carbon gas could be liquefied with its cooling, independently of the applied pressure. During the years 1850 and 1851, William Thompson and Rudolf Clausius, respectively, articulated the first and the second law of thermodynamics. The first patent for a refrigerating system was deposited in 1951 by a doctor named John Gorrie. These and other discoveries were crucial to the understanding of the behavior and characteristics of the gases, still formed the basis for the separation of gases developed in the following years (Yampolskii and Freeman 2010). In 1983, Bird presents the first results, involving carbon molecular sieve membranes, of the first measures of mass transfer and fluid gaseous diffusion through membranes. In 1985, William Schell presented in his work using gas separation membranes permeability and selectivity results of two different membranes of dimethyl silicone and other cellulose acetate. Still, it was found that the numerous advantages presented as high flux and selectivity in comparison with conventional methods (Schell 1985). These studies, as well as others, were pioneers in the area, and initiated a new application of the membrane technology.

2.6.2 Gas Separation Process

Gas mixtures can be separated by dense or porous ceramic membranes (Li 2007). Several researches have shown, effectively, the use of ceramic membranes in separation of gas mixtures. Ceramic membranes used in the separation of gases can be silica (Nwogu et al. 2016), kaolinite (Hubadillah et al. 2016), zeolite (Chew and Ahmad 2016; Kida et al. 2017), alumina (Donelson et al. 2014), Titania, zirconia

(Li and Qi 2015), and mullite (Ahmad et al. 2016). The dense ceramic membranes are made of crystalline ceramic materials, such as perovskite or fluorite, which allow only oxygen or hydrogen permeation through your crystalline structure. So are mostly impermeable to all other gases, giving extremely high selectivity for oxygen or hydrogen (Li 2007).

The quantitative measurement of the gas transport is the flow or the rate of permeation, which is defined as the number of molecules that pass through a unit area per unit time. The flow (*J*) follows the law of Fick, Eq. 2.16, and is proportional to the concentration gradient across the membrane, that is, there is a movement of gas from regions of high concentration to areas of low concentration.

$$J = -D\frac{\mathrm{d}C}{\mathrm{d}X} \tag{2.16}$$

$$J = D\frac{C_2 - C_1}{L} \tag{2.17}$$

where D is the diffusivity, $C(X)$ is the concentration, X is the position within the structure of the membrane, $C_1 = c\ (0)$, and $C_2 = C\ (L)$ and L is the thickness of the membrane (cm). Assuming a linear concentration gradient across the membrane, the flow can be approximated by Eq. 2.17.

The membrane performance of various materials is commonly compared by using the independent thickness material, the permeability P, which is related to the flux (*J*), Eq. 2.18.

$$P = \frac{Q \cdot L}{A \cdot (p_2 - p_1)} = \left(\frac{C_2 - C_1}{p_2 - p_1}\right) \cdot D \tag{2.18}$$

where P is the permeability in Barrer (1 Barrer = 10^{-10} cm^3 (STP) cm cm^{-2} s^{-1} cmHg), Q is the permeate flow rate (cm^3 s^{-1}), A is the area of the membrane (cm^2) and p_1 and p_2 are the pressures (cmHg) permeate side and food, respectively, Fig. 2.6.

If the upstream pressure (p_2) is much larger in relation to the downstream pressure (p_1) the Eq. 2.18 can be simplified in Eq. 2.19.

$$P = \frac{C_2}{p_2} \cdot D \tag{2.19}$$

$$P = S \cdot D \tag{2.20}$$

By introducing a solubility coefficient S, the concentration ratio on C_2/p_2 pressure when the sorption can be represented by Henry, Eq. 2.20.

The separation of a mixture of molecules A and B is characterized by selectivity or ideal separation factor, and can be calculated by the Eq. 2.21. The selectivity is represented by the ratio of the permeability of the component by component B permeability.

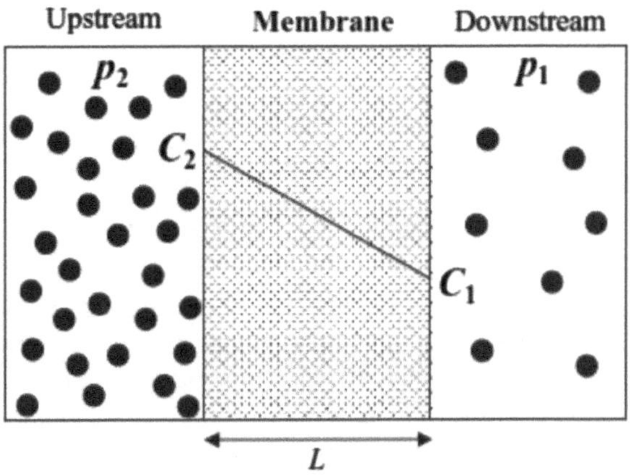

Fig. 2.6 Representation of a gas separation membrane with a constant concentration gradient through the thickness of the membrane (Yampolskii and Freeman 2010)

$$\alpha(\mathrm{A/B}) = \frac{P(\mathrm{A})}{P(\mathrm{B})} \tag{2.21}$$

Currently, several studies indicate the ceramic membranes as alternative hydrogen purification. Hydrogen has been an alternative to fossil fuels because it is considered a green technology, which can in future represent a considerable reduction of the air pollution degree. However, the main difficulty in separating hydrogen from other gases like N_2, CO_2 and CH_4, it is own very small dimension of atoms of gases, as well as in the proximity between the different gas molecules. For the separation to be effective it is necessary to have a control of porosity of selective layer or use of mechanisms that favor the separation. Because of this, some jobs present alternatively the use of palladium as a catalyst of hydrogen dissociation. However, the cost and complexity still are high. On the other hand, these difficulties have led the study of recovery of ceramic membranes used to support selective layers of palladium (Hu et al. 2011; Li et al. 2015).

2.7 Membrane Distillation

2.7.1 Membrane Distillation History

The early history of the Membrane Distillation (MD) is marked with your first patent deposited in 1963, by Bodell. In 1967, with the work of Findley, it was possible to access the first MD. Then, for about a decade, then was a lack of interest of researchers by MD, because there was no suitable membranes still for this process. The process of MD gained momentum again in the 80, when the first porous membrane of polytetrafluoroethylene (PTFE) began to be available on the market. Over the years, until 2013, there was an increase of approximately twenty times the number of publications, and in 2016, the number of publications was 442, which is almost five times more compared to the year 2013 (Khayet 2011; Basile et al. 2015).

2.7.2 *Membrane Distillation Process*

Most of the transports that occur in membranes is isothermal and its driving forces are differences in pressures, concentrations, electrical or chemical potentials. In simplified form, the MD is a thermal separation process that distills water. The process of MD is similar to conventional distillation, since both technologies are based on steam/liquid equilibrium for separation. Both processes require the use of heat in feed current. In the process, use a non-isothermal membrane with hydrophobic characteristics, porous, and this is not wet by the liquid streams. The steam heat transport through the membrane occurs by a difference in steam pressure (driving force) between the two sides of the porous membrane. The membrane is maintained in direct contact with the hot liquid feed solution to be treated. In the MD process occurs, at the same time, the mass and heat transfer (Khayet 2011; Alkhudhiri et al. 2012; Drioli et al. 2015; Basile et al. 2015). The MD can have different settings, namely:

Direct Contact MD (DCMD)
In this configuration, the difference in vapor pressure is achieved through a temperature difference between the two sides of the membrane, permeated and feed. As the cold liquid (permeated) flows from one side of the membrane, in order to condense the steam that runs through the pores of the membrane, the hot liquid (feed current) flows to the other side of the membrane. This setting can present high flows, but occurs a high loss in heat, and it provides a low efficiency (Hwang et al. 2011).

Air Gap MD (AGMD)
In the AGMD process, a gap of stagnant air is maintained between the membrane and the channel of the condenser using a condenser foil. The air gap acts as a layer of insulation. As a result, the heat loss due to driving, that is intrinsic to the DCMD, is noticeably reduced in this configuration. You can also separate the chains of volatile compounds even before mixing with the permeated. In addition, the AGMD enables retrieval of latent heat without the presence of an external heat exchanger. However, the existing barrier (air gap) causes a reduction of mass transfer through membrane (Duong et al. 2016).

Vacuum MD (VMD)
The VMD contains a setting that has a channel with an air gap. The membrane is used as a surface for the water to be evaporated in a vacuum environment,

maintained by steam absorption by a desiccator. The advantage of this configuration is that the inert gases do not dissolved, which block the pores, are aspirated by vacuum, leaving the effective membrane area available. Entertaining, the cost of the vacuum equipment and your adaptation to the system can cost a higher price (Chen et al. 2015).

Sweeping Gas MD (SGMD)

The SGMD, also known as air stripping, uses a configuration with a channel with empty space permeate side. The SGMD process is mainly used in removing volatile compounds. The feed solution is heated to the desired temperature and transferred to the surface of the membrane. While going through the pores in the direction of the permeated, volatile compounds are dragged by a gas transporter, which leads them to the further processing (extraction). Meanwhile, the non-volatile compounds remain recirculating in the feed. The advantage of this configuration in relation to AGMD is the reduction of additional mass transfer barrier and so could be achieved a higher flow. However, the gas flow can reduce the driving force (steam gradient across the membrane) due to the heating of the gas itself (Duyen et al. 2016).

Vacuum Multi-effect MD (V-MEMD)

The technology of memsys vacuum-multi-effect-membrane-distillate, as it is known, combines the advantages of multiple effect and vacuum to achieve efficient heat recovery. The V-MEMD is composed of a steam raiser, stages of evaporation-condensation and a capacitor. Each stage retrieves the heat of condensation, promoting a designer of multiple effect. The thermal energy produced by external heating source (for example, solar heat or other waste heat) is changed in steam raiser. The water from the evaporator is under pressure (for example, 600 mbar). The hot steam is generated by steam-raiser flows to the stage 1, while the feed current is also introduced in stage 1 and flows in series through the remaining stages. At the end of the last stage, the feed is concentrated as brine. The vacuum is always applied alongside permeated the membranes. The vapor pressure and temperature decrease gradually from steam raiser to the condenser. The steam produced in the final stage is condensed in the condenser, using a refrigerant flow (for example, fresh water) (Zhao et al. 2013).

DCMD configuration is the most studied, although the heat transferred by conduction through the membrane, considered the heat lost, is larger in relation the other settings. Almost 60% of the studies of the process of MD are conducted using DCMD systems. On the other hand, is a smaller configuration SGMD use (less than 5%), because external capacitor is required to collect the permeated, complicating the design of the system and increasing your cost (Basile et al. 2015).

One of the main features is the process of MD is the liquid entry pressure (LEP) of the membrane, and this must be greater than hydrostatic transmembrane pressure applied. The LEP is the minimum required for transmembrane pressure water or feed to enter the pore solution, surpassing the hydrophobic membrane forces. The LEP is related to the maximum pore size ($d_{p,\max}$) membranes by means of Laplace's equation, Eq. 2.22.

$$\mathrm{LEP} = \left|\frac{4.\alpha.\sigma}{\mathrm{d}_{p,\max}} \cdot \cos\theta\right| \tag{2.22}$$

where α is the geometric factor the greater the membrane pore, σ is the surface tension of the liquid solution and θ is the angle of contact between the liquid and the surface of the membrane.

A membrane with a value of LEP high can be developed using materials of low surface energy or with high hydrophobicity and small pore size. However, when a membrane displays a small pore size is expected to be a low permeability of the membrane. Besides, the size of your distribution and pore is considered crucial in the process of MD. The pore size can vary from 5 nm to 10 μm, however your distribution should be as close as possible.

The Knudsen number (Kn) is another key parameter in MD, and is used to determine the mass transport through membrane pore. Equation 2.23 presents the Knudsen number math definition, and is used under certain conditions.

$$Kn = \frac{\lambda i}{\mathrm{d}_p} \tag{2.23}$$

$$\lambda_i = \frac{k_B.T}{\sqrt{2\pi.P_m.\sigma_i^2}} \tag{2.24}$$

where λ_i is average free path of the molecules of steam carried through the pores of the membrane with a d_p size (Eq. 2.24), k_B is the Boltzmann constant, T is the absolute temperature, P_m is the average pressure within the pores of the membrane, and σ_i is the collision diameter.

When it comes to a binary mixture, the value of λ can be calculated using the Eq. 2.25.

$$\lambda_{i/j} = \frac{k_B.T}{\pi.P_m.((\sigma_i + \sigma_j)/2)^2} \cdot \frac{1}{\sqrt{1 + M_j/M_i}} \tag{2.25}$$

where σ_i and σ_j are the collision diameters and M_i and M_j are the molecular masses of the molecules i and j, respectively.

The Knudsen number is the basis to determine the permeability and permeate flow to various process conditions. Several studies present models to determine the permeability and the permeate stream, but here will not be dealt with these models. For more informations, it is recommended the consult of recommended literature (Basile et al. 2015).

The process of MD uses organic and inorganic membranes. Several studies using ceramic membranes composed mainly of zeolite, zirconia, alumina, and Titania (Fang et al. 2012; Kujawski et al. 2016; Garofalo et al. 2016; Fan et al. 2017). The various advantages of ceramic membranes with polymer make them interesting in the process of MD.

References

Ahmad, R., Anwar, M.S., Kim, J., Song, I., Abbas, S.Z., Ali, S.A., Ali, F., Ahmad, J., Awais, H., Mehmood, M.: Porosity features and gas permeability analysis of bi-modal porous alumina and mullite for filtration applications. Ceram. Int. **42**(16), 18711–18717 (2016)

Alkhudhiri, A., Darwish, N., Hilal, N.: Membrane distillation: a comprehensive review. Desalination **287**, 2–18 (2012)

Basile, A., Figoli, A., Khayet, M.: Pervaporation, Vapour Permeation and Membrane Distillation: Principles and Applications. Elsevier Science, Cambridge (2015)

Bird, A.J.: Carbon molecular sieves used in gas separation Membranes. Carbon **21**(3), 177–180 (1983)

Chaabane, T., Taha, S., Ahmed, M.T., Maachi, R., Dorange, G.: Coupled model of film theory and the Nernst–Planck equation in nanofiltration. Desalination **206**, 424–432 (2007)

Chen, Z., Rana, D., Matsuura, T., Meng, D., Lan, C.Q.: Study on structure and vacuum membrane distillation performance of PVDF membranes: II. Influence of molecular weight. Chem. Eng. J. **276**, 174–184 (2015)

Cheryan, M.: Ultrafiltration and Microfiltration Handbok. Technomic Publishing Company, Basel (1998)

Chew, T.L., Ahmad, A.L.: Gas Permeation Properties of Modified SAPO-34 Zeolite Membranes. Proc. Eng. **148**, 1225–1231 (2016)

Donelson, R., Paul, G., Ciacchi, F., Badwal, S.: Permeation and strength characteristics of macroporous supports for gas separation produced by co-sintering mixtures of α-alumina and kaolin. J. Membr. Sci. **463**, 126–133 (2014)

Drioli, E., Ali, A., Macedonio, F.: Membrane distillation: recent developments and perspectives. Desalination **356**, 56–84 (2015)

Duong, H.C., Cooper, P., Nelemans, B., Cath, T.Y., Nghiem, L.D.: Evaluating energy consumption of air gap membrane distillation for seawater desalination at pilot scale level. Sep. Purif. Technol. **166**, 55–62 (2016)

Duyen, P.M., Jacob, P., Rattanaoudom, R., Visvanathan, C.: Feasibility of sweeping gas membrane distillation on concentrating triethylene glycol from waste streams. Chem. Eng. Process. **110**, 225–234 (2016)

Fan, Y., Chen, S., Zhao, H., Liu, Y.: Distillation membrane constructed by TiO_2 nanofiber followed by fluorination for excellent water desalination performance. Desalination **405**, 51–58 (2017)

Fang, H., Gao, J.F., Wang, H.T., Chen, C.S.: Hydrophobic porous alumina hollow fiber for water desalination via membrane distillation process. J. Membr. Sci. **403–404**, 41–46 (2012)

Garofalo, A., Carnevale, M.C., Donato, L., Drioli, E., Alharbi, O., Aljlil, S.A., Criscuoli, A., Algieri, C.: Scale-up of MFI zeolite membranes for desalination by vacuum membrane distillation. Desalination **397**, 205–212 (2016)

Gitis, V., Rothenberg, G.: Ceramic Membranes: New Opportunities and Practical Applications. Wiley-VCH Verlag GmbH & Co. KgaA, Weinheim (2016)

Hilal, N., Al-Zoubi, H., Darwish, N.A., Mohammad, A.W., Arabi, M.A.: A comprehensive review of nanofiltration membranes: treatment, pretreatment, modelling, and atomic force microscopy. Desalination **170**, 281–308 (2004)

Ho, W.S.W., Sirkar, K.: Membrane Handbook. Van Nostrand Reinhold, New York (1992)
Hu, X., Yu, J., Song, J., Wang, X., Huang, Y.: Toward low-cost Pd/ceramic composite membranes for hydrogen separation: a case study on reuse of the recycled porous Al2O3 substrates in membrane fabrication. Int. J. Hydrogen Energy **36**, 15794–15802 (2011)
Hubadillah, S.K., Harun, Z., Othman, M.H.D., Ismail, A.F., Salleh, W.N.W., Basri, H., Yunos, M. Z., Gani, P.: Preparation and characterization of low cost porous ceramic membrane support from kaolin using phase inversion/sintering technique for gas separation: Effect of kaolin content and non-solvent coagulant bath. Chem. Eng. Res. Des. **112**, 24–35 (2016)
Hwang, H.J., He, K., Gray, S., Zhang, J., Moon, I.S.: Direct contact membrane distillation (DCMD): Experimental study on the commercial PTFE membrane and modeling. J. Membr. Sci. **371**(1–2), 90–98 (2011)
Khayet, M.: Membranes and theoretical modeling of membrane distillation: a review. Adv. Colloid Interface Sci. **164**, 56–88 (2011)
Kida, K., Maeta, Y., Yogo, K.: Preparation and gas permeation properties on pure silica CHA-type zeolite membranes. J. Membr. Sci. **522**, 363–370 (2017)
Kujawski, W., Kujawa, J., Wierzbowska, E., Cerneaux, S., Bryjak, M., Kujawski, J.: Influence of hydrophobization conditions and ceramic membranes pore size on their properties in vacuum membrane distillation of water–organic solvent mixtures. J. Membr. Sci. **499**, 442–451 (2016)
Lee, K.P., Arnot, T.C., Mattia, D.: A review of reverse osmosis membrane materials for desalination—development to date and future potential. J. Membr. Sci. **370**(1–2), 1–22 (2011)
Li, K.: Ceramic Membranes for Separation and Reaction. Wiley, London (2007)
Li, L., Dong, J., Nenoff, T.M., Lee, R.: Desalination by reverse osmosis using MFI zeolite membranes. J. Membr. Sci. **243**(1–2), 401–404 (2004)
Li, L., Qi, H.: Gas separation using sol–gel derived microporous zirconia membranes with high hydrothermal stability. Chin. J. Chem. Eng. **23**(8), 1300–1306 (2015)
Li, Y., Ding, W., Jin, X., Yu, J., Hu, X., Huang, Y.: Toward extensive application of Pd/ceramic membranes for hydrogen separation: a case study on membrane recycling and reuse in the fabrication of new membranes. Int. J. Hydrogen Energy **40**, 3528–3537 (2015)
Loeb, S., Sourirajan, S.: Sea water demineralization by means of an osmotic membrane. Am. Chem. Soc. **38**, 117–132 (1962)
Metcalf and Eddy: Wastewater Engineering: Treatment and Resource Recovery. McGraw-Hill Education, Boston (2014)
Meyer, P., Meyer, A., Kulozik, U.: High concentration of skim milk proteins by ultrafiltration: Characterization of a dynamic membrane system with a rotating membrane in comparison with a spiral wound membrane. Int. Dairy J. **51**, 75–83 (2015)
Mohammad, A.W., Teow, Y.H., Ang, W.L., Chung, Y.T., Oatley-Radcliffe, D.L., Hilal, N.: Nanofiltration membranes review: Recent advances and future prospects. Desalination **356**, 226–254 (2015)
Nwogu, N.C., Anyanwu, E.E., Gobina, E.: An initial investigation of a nano-composite silica ceramic membrane for hydrogen gas separation and purification. Int. J. Hydrogen Energy **41** (19–25), 8228–8235 (2016)
Oyama, S.T., Stagg-Williams, S.M.: Inorganic, Polymeric and Composite Membranes: Structure, Function and other Correlations. Elsevier Science, Oxford (2011)
Paul, M., Jons, S.D.: Chemistry and fabrication of polymeric nanofiltration membranes: a review. Polymer **103**, 417–456 (2016)
Reid, C., Breton, E.: Water and ion flow across cellulosic membranes. J. Appl. Polym. Sci. **1**(2), 133–143 (1959)
Schell, W.J.: Commercial applications for gas permeation membrane systems. J. Membr. Sci. **22**, 217–224 (1985)
Scott, K.: Handbook of Industrial Membranes. Elsevier Science, Oxford (1998)
Singh, R.: Hybrid Membrane Systems for Water Purification: Technology, Systems Design and Operations. Elsevier Science, Oxford (2006)

Wang, L.K., Chen, J.P., Hung, Y.-T., Shammas, N.K.: Handbook of Environmental Engineering: Membrane and Desalination Technologies. Springer Science + Business Media, London (2011)

Yampolskii, Y., Freeman, B.: Membrane Gas Separation. Wiley, West Sussex (2010)

Zhao, K., Heinzl, W., Wenzel, M., Büttner, S., Bollen, F., Lange, G., Heinzl, S., Sarda, N.: Experimental study of the memsys vacuum-multi-effect-membrane-distillation (V-MEMD) module. Desalination **323**, 150–160 (2013)

Chapter 3
Ceramic Membranes Preparation

In this chapter, emphasis will be given to the main raw materials used in production of ceramic membranes as well as with the main processes of obtaining the same. The need to understand the phenomena involved in the processing of ceramic membranes, as well as the interaction of materials used is of utter importance for the performance of the final product, in accordance with the type of process used.

3.1 Raw Materials Used in the Preparation of Ceramic Membranes

Ceramic membranes are generally composed of several layers of one or more different ceramic materials. These layers are divided into a support layer (Macropores), in one or more intermediate layers (mesoporous), and in a filtering layer (microporous or dense), Fig. 3.1.

The lower layer provides mechanical support, while the intermediate layers gather the pores between the support layer and the selective layer, where the separation occurs. Ceramic membranes can be synthesized and drawn from a variety of materials, and they can be titania, alumina, silica, and among others. These membranes are often used in harsh environments and high temperatures due to their good thermal and chemical stability. Given this, the main raw materials used in the manufacture of ceramic membranes.

3.1.1 Alumina

Aluminum oxide (Al_2O_3), alumina, is the most widely used ceramic material in microelectronics industry due to its resistance to abrasion, wear, high thermal

D. da Silva Biron et al., *Ceramic Membranes Applied in Separation Processes*,
Topics in Mining, Metallurgy and Materials Engineering,
DOI 10.1007/978-3-319-58604-5_3

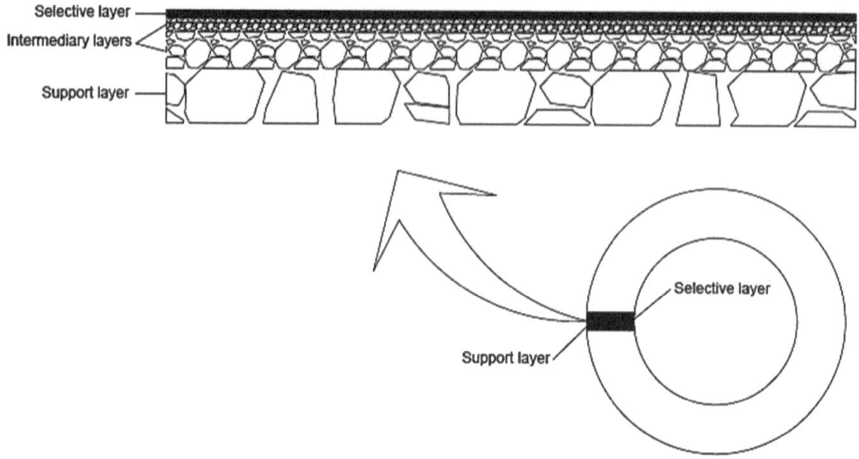

Fig. 3.1 Representation of the structure, in layers, of tubular ceramic membranes

stability, and high electrical resistance are superior to other ceramic materials. The raw material is abundant, has low cost, and can be manufactured in a variety of techniques available and be composed in a variety of formats. Various other applications of alumina can be highlighted: as, polishing powder material for crucibles, cutting tools, spark plugs, substrate for electronic circuits, to the hip joints, as well as a filter for the water and wastewater treatment. Alumina is the material most commonly associated with the ceramic membranes (Riedel 2000; Gitis and Rothenberg 2016).

The alumina is produced from bauxite-treated primarily through the Bayer process. Aluminum production consumes about 85% of bauxite and 10–15%, the bauxite is consumed for alumina production and up to 5% is used in the form of bauxite calcined. Bauxite is a combination of aluminum hydroxide mixtures containing, according to location, various impurities such as SiO_2, TiO_2, and FeO_3. The chemical purity of alumina will depend on the process. Alumina obtained by the Bayer process typically contains 0.3–0.5% of Na_2O, less than 0.1% SiO_2, and less than 0.03% other nonvolatile impurities. The presence of Na_2O is due to the alkaline decomposition process and its quantity depends on the washing process. The properties of alumina, particle size and shape, surface area specifies and reactivity, depend on, particularly, the calcination temperature (up to about 1400 ° C), the type of initial hydroxide, and structural transformation sequence (Hlavác 1983; Boch and Nièpce 2007).

Alumina has several meta phases, including γ-alumina and the α-alumina, the two main forms used in the preparation of membranes. High purity alumina, 99.99% or more, in forms of ultrathin powders can be synthesized by advanced processes: hydrolysis and heat treatment of aluminium alkoxide and pyrolysis of ammonium aluminium, $(NH_4)_2SO_4{\cdot}Al_2(SO_4)_3{\cdot}12H_2O$, or ammonium dawsonite, $NH_4AlCO_3(OH)_2$, through the following reactions (Somiya 2003):

- Hydrolysis and thermal treatment of the alkoxide of alumina
  $2Al + 6ROH \rightarrow 2Al(OR)_3 + 3H_2$
  $2Al(OR)_3 + 4H_2O \rightarrow Al_2O_3 \cdot H_2O + 6ROH$
  $Al_2O_3 \cdot H_2O \rightarrow \text{calcination} \rightarrow Al_2O_3 + H_2O$
- Pyrolysis of ammonium aluminium
  $Al_2(SO_4)_3 \cdot 18H_2O + (NH_4)_2SO_4 \rightarrow (NH_4)_2SO_4 \cdot Al_2(SO_4)_3 \cdot 12H_2O + 6H_2O$
  $(NH_4)_2SO_4 \cdot Al_2(SO_4)_3 \cdot 12H_2O \rightarrow \text{pyrolysis} \rightarrow Al_2O_3 + 2NH_3 + 4SO_3 + 13H_2O$
- Pyrolysis of ammonium dawsonite
  $(\text{Aluminum salt.}) + NH_4HCO_3 \rightarrow NH_4AlCO_3(OH)_2$
  $2NH_4AlCO_3(OH)_2 \rightarrow Al_2O_3 + 2NH_3 + 2CO_2 + 3H_2O$

Powders manufactured by these methods exhibit characteristics such as high purity, particles with sizes below 1 μm and uniform size distribution (Somiya 2003).

The α-alumina is considered stable alumina phase, in other words, the final stage (Somiya 2003; Boch and Niepce 2007). Some authors have studied the intermediate transition of alumina phases (Pijolat et al. 1987; Santos et al. 2000; Xu et al. 2017) and according to them the nature and size of the particles, the activity of the material, the presence of gas, impurities and additives in calcination atmosphere can cause the growth of crystals or affect other properties, as well as the routes and conditions of preparation and thermal treatments. These characteristics may influence the sequence of transition, as well as on phase of aluminum oxide formed (Gitzen 1970).

3.1.2 *Zeolites*

The discovery of the zeolite was around the year 1756 by a swedish mineralogist called Axel Fredrik Cronstedt. He discovered that burning the mineral stilbite, this seemed to boil. The new family of microporous crystalline aluminosilicates materials takes its name from the Greek words zeolites zeo and lithos meaning boil and stone, respectively. These materials and microporous crystalline aluminosilicates consist of a three-dimensional arrangement tetrahedral TO_4 (SiO_4 or AlO_4^-) linked together to form subunits and, finally, huge polymeric networks consisting of identical blocks, which gives the name of unit cells. Currently are listed over 100 types of structures presented by zeolites, and around 40 structures can be found in nature. The zeolite structure in its regular three-dimensional pores and is relatively stable at high temperatures. As zeolites are currently used as catalysts or catalyst carriers for a series of reactions at high temperatures. As naturally occurring zeolites of hydrothermal alteration of volcanic lava, it is the influence of water at high temperatures. This process occurs over years, supplies make material a decorating object exposed on Rails for 200 years, as its beautiful and distinct forms. With the discovery of zeolites, their discoveries and applications have emerged in several areas (Szostak 1989; Li 2007).

The zeolites are crystalline oxides and are different from other materials due to its unique properties, as for example the microporous pore dimensions features character uniforms, allowing certain molecules of hydrocarbons entering the crystals while rejecting other with larger molecular, ion-exchange properties that allow one to perform all kinds of ion-exchange reactions, has ability to develop an internal acidity that make them interesting materials to catalyze organic reactions and has high thermal stability (Bekkum et al. 1991; Auerbach et al. 2003).

The polycrystalline zeolite membranes' preparation started in the late 80, and were the MFI-type zeolite membranes (family of zeolites with the structure containing pentasil units) that presented interesting properties of permeation and separation. Since then, the synthesis of zeolite membranes became a very active research field and a considerable number of research articles were published until then. It is expected that a zeolite membrane behaves like a solid separation device, with selectivity and reasonably high flow. Many of the separations that are currently carried out using distillation, crystallization, or other conventional processes can be performed with a separation process based on zeolite membrane. Advantage there would be a reduction of the energy costs (Li 2007).

According to Auerbach et al. (2003) there are two steps to be considered in a zeolite membrane synthesis. The first is to identify a zeolite material expected to possess the ability to perform a particular type of separation. This information is usually derived from adsorption and data obtained using diffusivity powder samples or single crystals, and indicates an estimated level of permeability and selectivity that can be achieved with the zeolite. In the second, it is desirable to have bibliographic information on the chemical synthesis of zeolite, i.e., the temperature required for hydrothermal synthesis and reagents required. The agents include a source of silica and an alkali hydroxide (NaOH, KOH); still, a structural direction (ADS) and an organic source of alumina. On the basis of this information, it a method is necessary for preparation of these membranes. According to Li (2007), there are basically two approaches to synthesize zeolite membranes: synthesis in liquid phase and the phase transport process steam (TFV). Liquid phase synthesis consists in placing the porous support surface in contact with a solution of zeolite synthesis (sol or gel) and keep the system under controlled conditions to zeolitic material can grow and nuclear to a continuous film on the surface of the support. The method of transport in steam phase treats a gel precursor deposited on a stand with dry saturated steam at a temperature similar (Auerbach et al. 2003; Li 2007).

In the literature, there are a number of studies with composite zeolite-polymer membranes. These membranes are characteristic to develop materials with desirable properties and pervaporation separation processes. According to Chakraborty et al. (2016), composite polymer–zeolite membranes possess greater flexibility regarding composite membranes of alumina–zeolite and their mechanical properties depend on your elastic constants. It is possible to obtain polyethersulfone -zeolite membranes from the nucleation and growth of zeolite nanoparticles on the surface of polyethersulfone films. The amount of zeolite nanoparticles significantly influences the mechanical properties of the membrane, promoting an increase in ductility with increasing concentration of zeolites and consequently produces a greater flexibility of the membrane (Wang et al. 2015; Chakraborty et al. 2016).

3.1.3 *Titania*

Titanium oxide (TiO_2), titania, is another popular material used for making ceramic membranes. Titania membranes feature excellent chemical resistance and can be used in acid and basic pH. In addition, they show photocatalytic activity interesting. The titania occurs in nature primarily in the form of ilmenite ore ($FeO{\cdot}TiO_2$) and in your composition can hold approximately 15–40% iron. Still, the titania can meet in the form of rutile ore and may contain in your composition until 15% iron and other impurities. The titania has three crystalline forms, and two, anatase and brookite are meta states and, after heating reaches your final stage, Auburn. The tetragonal rutile crystal structure possesses and is the only used commercially (Li 2007).

The titania may be used in obtaining membranes or serve as top layer. Titania membranes are produced in the form of tubes, may reach lengths in micrometric and nanometric scale. Titania of UF membranes have been marketed by several companies. Currently, efforts are concentrated on the preparation of porous membranes with pore sizes in the nanoscale. Tsuru et al. were the first to get applied membranes in separation of organic compounds, the authors obtained with titania membranes in their composition, by sol–gel method, with pore sizes in the nanoscale and the applied in filtering and permeation of organic solvents (Tsuru et al. 2000, 2001, 2008). The use of ceramic materials for membrane preparation aimed at the separation of organic solvents has great potential due to its advantages in respect of polymeric materials, such as resistance to compression and to swelling. However, the high costs of ceramic membranes and low rejection of organic solvents limit your application in separation of organic solvents. In this way, the research has conducted surface modification techniques and functionalization to adjust the hydrophilic and hydrophobic characteristics of NF ceramic membranes and, consequently, improve the performance of your separation (Tsure et al. 2000, 2001, 2008; Amirilargani et al. 2016).

3.1.4 *Zirconia*

The zirconium is widely distributed in the Earth's crust. It is found in crystalline rocks (especially limestone), gneiss, syenites, granite and beds of iron ore, pegmatite, and ferruginous sandstone. It is more abundant than copper, lead, tin, zinc, nickel, and other base metals. It is usually found as an oxide or silicate. Due to the chemical similarity between zirconium (Zr) and hafnium (Hf) these occur together in nature. The zirconium mineral Baddeleyite is best known, as zirconium or zircon and caldasito. Zirconium dioxide (ZrO_2), or zirconia, is found naturally in the form of Baddeleyite, but is most often prepared from zircon (ZrSiO) by high-temperature heat treatments, accompanied by chemical treatments that eliminate the fraction of siliceous zircon (Boch and Nièpce 2007).

The zirconia features applications such as refractory material, and can work with temperatures greater than 2000 °C in oxidizing atmosphere, and in the manufacture of dyes, abrasive material, opacifiers for glazes and glazed over and in the manufacture of dielectric porcelain. Also, for being stable in alkaline solution, as well as the titania, zirconia is a promising material for separation membranes, primarily as ultrafiltration membranes (Gitis and Rothenberg 2016).

The zirconia features three crystalline forms, monoclinic, tetragonal, and cubic. The polymorphic form of monoclinic is more studied due to your stability to your temperature and ambient pressure. The change in crystalline form to another occurs with increasing temperature. Phases of transition from zirconia occur in the following order: tetragonal to monoclinic occurs approximately 1173 °C and the cubic 2370 °C. One of the characteristics of zirconia membranes is these can suffer cracks at high temperatures, known fact due to a large variation in volume (around 9%) in the tetragonal phase transition to monoclinic, during cooling. To prevent this, the Zirconia can be doped with yttria (Y_2O_3), calcium oxide (CaO), magnesium oxide (MgO) and among others, in its cubic Polymorph, which increases your thermal stability (Boch and Nièpce 2007).

3.1.5 Silica

Silica, or silicon dioxide (SiO_2), is widespread in nature in its pure form, so much so that in silicate technology that ends up being used in its raw form. Silica has a tetrahedral structure in which each silicon atom surrounded by four oxygen atoms. The SiO_4 has different polymorphic forms, the basic shapes include quartz ore, cristobalite, and tridymite. β-quartz is considered the polymorphic form more stable (below 573 °C), followed by β-quartz (among the 867,573 °C), tridymite β2 (between 867 the 1470 °C), cristobalite α (1470 the 1710 °C) and liquid (above 1710 °C) (Kingery et al. 1976; Gitis and Rothenberg 2016).

Silica can be applied in the production of glass, electrical and electronic equipment in construction, refractories, and among others. Still, the silica has proved a material that offers interesting features for making ceramic membranes, and can these be applied in separation of gaseous and liquid fluids. The biggest disadvantage presented by silica is your hydrothermal instability (Li et al. 2015). In general, the ceramic membranes in your silica structure presents a lifetime of a few days, even with moderate temperatures of up to 100 °C operation. Moisture and heat causes the physical adsorption of water molecules to the Si–OH group of the silanol followed by a reaction with Group links siloxane (Si–O–Si) (Gitis and Rothenberg 2016).

In the study of Peyki et al. (2015) the authors added silica nanoparticles-shaped reverse osmosis membranes of polyamide aiming at the modification of the hydrophilicity surface membrane properties. The addition of silica promoted the increase of roughness and of hydrophilic and anti-fouling properties of the membrane. These features are crucial to the performance of membranes applied in desalination processes. The fouling has the main responsibility for the loss of quality of the membrane, causing increased energy consumption and operating costs. However, the modification of surface properties of the membranes with silica is a way to solve or minimize the incidence of fouling (Peyki et al. 2015).

In general, the methods used to obtain silica membranes are the sol–gel and method of chemical steam deposition. The sol–gel method is the oldest method for inorganic porous membranes, which usually includes immersion coating of silica precursors in a porous support, followed by controlled drying and calcining at elevated temperature. The chemical steam deposition method consists in the thermal deposition precursors, followed by chemical reaction with an oxidizing agent. The membranes are usually amorphous silica and are more easily made in a thin layer ultra-microporosa which displays an excellent property of molecular separation (Li et al. 2015).

3.1.6 *Other Materials*

In addition to the above, many other materials can be used to obtain ceramic membranes. It is possible to obtain carbon membranes, mullite, cordierite mullite, hafnia, silicon carbide, and among others. The properties of these materials arouse your interest in the constant application in membrane separation processes. For example, carbon membranes are highly selective due to the pores of molecular dimensions. These membranes are relatively inexpensive and are prepared basically for carbonization of organic polymers, as starting materials, high temperatures under controlled conditions. Another example is the mineral mullite, which is the only stable crystalline intermediate phase in Al_2O_3–SiO_2 binary system in the range of 70.5–74.0% in weight of Al_2O_3–SiO_2 at atmospheric pressure. In recent years, the increasing applications in the field had mullite ceramics, structural and functional, in response to its excellent physical properties such as low thermal expansion, low thermal conductivity, high melting point, and low density.

3.2 Processes Applied in Ceramic Membrane Preparation

The preparation of ceramic membranes has received much attention in recent decades. To get the best efficiency of operation of these materials requires a strict control over the different stages of manufacture of membrane. The key factor is a thorough understanding of the relationship between processing parameters, membrane structure, and properties. The following will be discussing some of the most important processes of preparation of ceramic membranes.

3.2.1 Extrusion

Extrusion has been used for the manufacture of ceramic materials for more than 150 years, and the technology involved suffered minor changes since 1950 (Blackburn and Lawson 1992). The meaning of extruding is to force the passage of a given material through a stoma. To best illustrate, imagine a medicinal ointment tube being expressed, in this case, the ointment contained in the tube is being extruded out. Therefore, the extrusion is a process where the material, along with additives such as binders, plasticisers, lubricants, are forced to go through an array, resulting in a material with a desirable format, Fig. 3.2 (Rauwendaal 1998).

The extrusion is one of the most used techniques in preparation of ceramic membranes, and can she produce a large quantity (over 1 ton) or small quantities (a few grams) of material (Li 2007). In general, this type of processing is applied to ceramic products with constant cross section.

To achieve good results in the extrusion of ceramic materials is necessary to have a ceramic mass with few variations in your composition, degree of milling and moisture content. The extrusion process is similar to the process fiber-spinning, however there are some differences. In the extrusion process, a folder is compressed and shaped, which is forced through a nozzle. One of the requirements in the

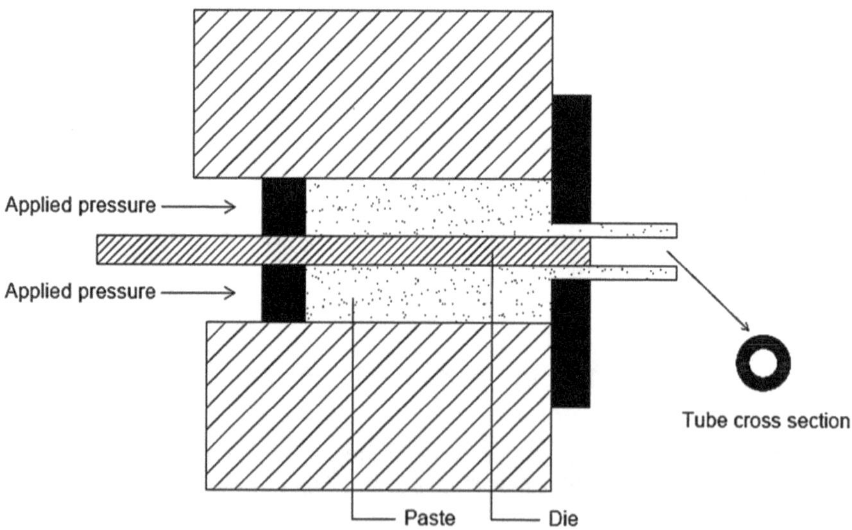

Fig. 3.2 Simplified process of extrusion

process of extrusion is that the forerunner should display plastic behavior. In the process, fiber-spinning, a viscous solution or suspension is transformed into a stable coagulation bath through a die. In addition, the forerunner made by extrusion has a homogeneous structure in your cross section, on the other hand, in the process of fiber-spinning the structure is asymmetric (Richerson 2006).

There are several factors that influence the performance of the extrusion process and the quality of the product. The manufacturer of ceramic materials that use this type of procedure should be concerned not only with the preparation, composition, and plasticity of materials, but also with the extruder type that will be used and the quality of the vacuum. Among the factors mentioned, plasticity is a property of hard and characterization will depend on several aspects, such as moisture content, composition, shape of the crystals, and granulometry of particles involved (Rice 2003; Bauser et al. 2006).

There are two types of extruders used in processing of ceramic materials which are the screw and piston. Yet, we can have several variations on the basis of these two types. In a piston-type extruder, material is forced forward and passes through an orifice, Fig. 3.2. This type of extruder presents good characteristics of transportation and can develop high pressures. The disadvantage of this extruder is its low melting capacity and homogenization. Conversely, the screw-type extruder is characterized by a cylinder that it has function rotate and push the ceramic mass, being driven by the action of a pump. Its advantage is its ability to merge and mix the material that's being extruded (Rauwendaal 1998).

3.2.2 *Slip Casting*

The slip casting is considered to be one of the most used methods in preparation of membranes. Most of the processes of slip casting involve ceramic particles suspended in water and merged into porous plaster molds. One of the problems surrounding this technique is the difficulty to control the thickness of the wall that it gets thicker. Figure 3.3 features the traditional method employed in slip casting process. In this, we observe a suspension powder mixed being poured into a porous mold, the suspension is solvent extracted through the pores of the mold, using the driving force of capillary suction. Slip particles are consolidated on the surface of the porous mold, forming a gel layer. It is important that the consolidation of layer occurs quickly to prevent the particles penetrate the pores of the mold (Li 2007; Richerson 2006).

Figure 3.4 shows all critical stages of the process slip casting and some of the parameters that need to be carefully controlled. The slip casting can be understood in the following steps:

(a) preparing the slip;
(b) preparation of mold;

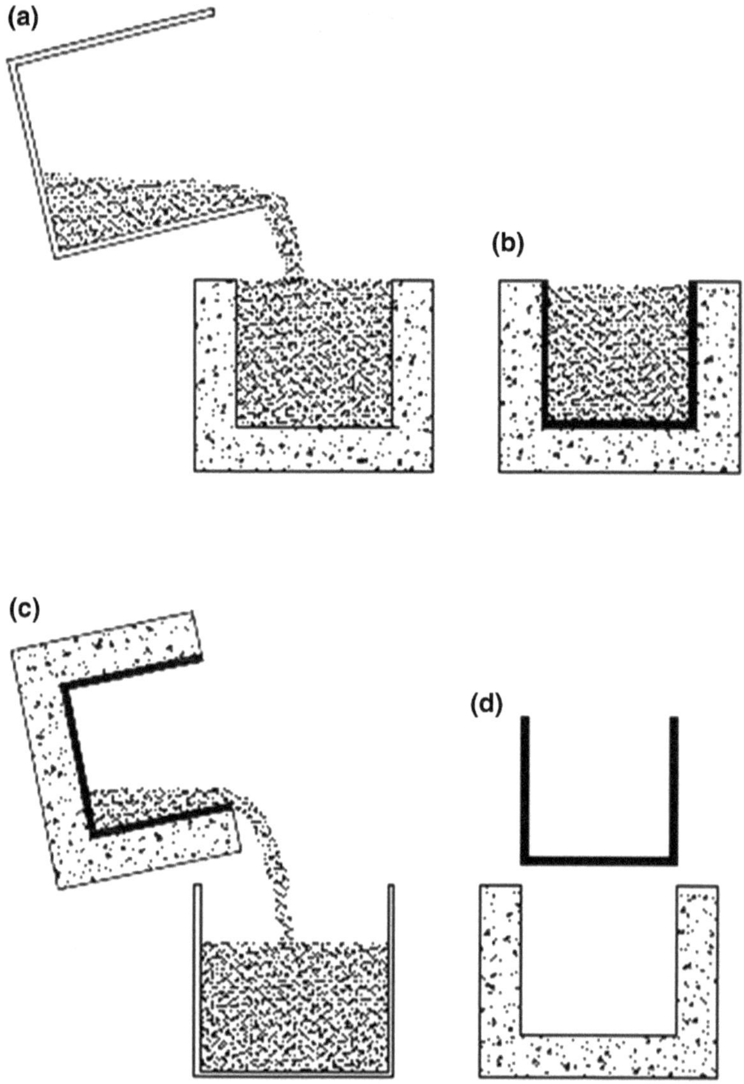

Fig. 3.3 Simplified slip casting process. **a** Filling the mold, **b** absorbing the fluid through the pores of the mold, **c** removing the exceeded paste, and **d** removing the formed piece

(c) molding (casting);
(d) process control of casting;
(e) drying.

The selection of initial powder depends on the type of application. Most applications require that the powder used be thin, typically 325 mesh, which equals

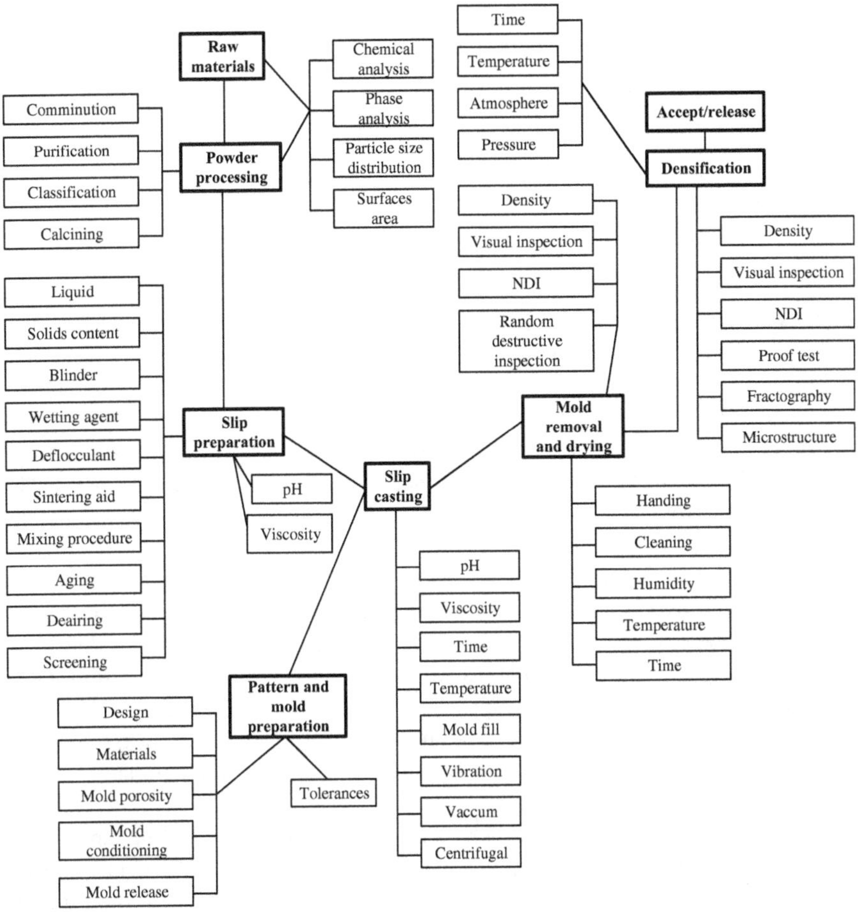

Fig. 3.4 Critical steps in the process of slip casting (Richerson 2006)

44 μm. The slip can be prepared in various ways. The most common method is the wet ball milling. The ingredients include dust, binders, Humidifiers, sintering and dispersing agents. The mold should introduce controlled porosity and need to introduce low cost. The plaster mold is considered the most traditional, however some molds used is plastics be features. The kind of casting to be used will depend on the complexity of the components and other factors. During the casting process, some factors were to be carefully studied, among them we can mention, viscosity, shrinkage, resistance, drainage properties, presence of air bubbles and among others. Finally, last but not least, we have the drying process. In this step, it is important that fluid is completely removed before the material be taken at high temperatures for densification (Richerson 2006). The removal of the fluid will depend on a few factors:

- porosity;
- size of the interconnected pore channels;
- steam pressure of the fluid;
- thickness of the gel formed.

3.2.3 Tape Casting

The use of the method of tape casting to form ceramics occurred primarily during the Second World War, the development of dielectric materials as a replacement for mica for capacitors of high quality. Glenn Howatt has developed a patent describing the method and founded a company that is now part of the Gulon Industries (Onoda and Hench 1978).

The tape casting is a technique used for the manufacture of flat, thin ceramic pieces (0.02–1.3 mm) and with large surface areas (Onoda and Hench 1978). This technique has aroused growing interest, especially in applications related to the electronics industry.

The casting film is formed by the use of a razor blade scraping property (casting knife), which promotes a thin layer of uniform material on a surface (casting tape or glass). The tape casting is very similar to casting method used in the production of polymeric membranes of flat sheet, as well as to the method of slip casting. Other methods such as extrusion and compression for rolls, are used for the production of thin ceramic pieces. However, do not show satisfaction in the process by being limited, as it is not possible to reach easily dimensions smaller than 0.25 mm (Richerson 2006).

The process known as doctor blade is the most common within the tape casting method. This process consists in pouring a folder on a mobile surface and spread it in a controlled thickness with the aid of a long and smooth blade. At the time that the surface comes into movement occurs the spreading of paste and then there is the evaporation of the solvent and formation of the leaves. For better understanding, Fig. 3.5 shows the process doctor blade. As in slip casting, the fluid must be removed slowly to not occur cracks, bubbles or distortions in the final product. The evaporation of the solvent must be controlled using heating and/or air flow pre-set depending on the type of material to be prepared (Onoda and Hench 1978; Richerson 2006).

The process consists of a stationary casting knife or doctor blade, a reservoir to suspensions in powder, a move support and a zone of drying (Li 2007; Richerson 2006). In the preparation of ceramic membranes of flat sheets, the suspension in powder is poured to within of a reservoir behind the doctor blade and the support that receive the material to be have molded is collocated in moving. The interval between casting tape and doctor blade is what determines the thickness of the ceramic sheet. Other important variables include reservoir depth, velocity, and viscosity of the material to gain the shape of the leaves. The wet material layer

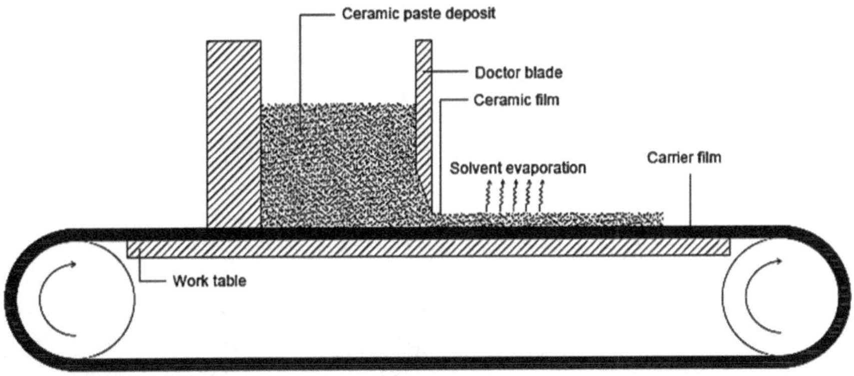

Fig. 3.5 Simplified representation of the process doctor blade

passes into a drying chamber, and the solvent is evaporated in contact with the surface, leaving a layer of the material (membrane) dried on the surface of the carrier (Li 2007).

3.2.4 *Dip-Coating*

Ceramic membranes are structured in various asymmetrical layers, and for one of them is aimed at separating layer function, where the process of retention of unwanted particles. In addition to this layer, the membrane has 1–5 layers responsible for the support of the assembly. This layer is responsible for the mechanical stability and must necessarily have greater porosity compared the separation layer (Burggraaf and Cot 1996).

Porous ceramic coating on porous substrates, can be realized by immersion of the substrate in a dispersion ceramic and then withdrawing it. This method is commonly called dip-coating and is often used to coat one or more layers of ceramics in a porous tubular support. Figure 3.6 shows simply the technique of dip-coating. The critical factors in dip-coating are the viscosity of the suspension of particles, the speed, and the finish time. The drying process occurs when the substrate is in contact with an atmosphere with a relative humidity of less than 100% (Burggraaf and Cot 1996; Li 2007).

3.2.5 *Pressing*

The pressing is a method that uses a particular pressure to compress and shape the granular powder, premixed with their additives, contained in an array or a flexible mold (Reed 1995). In general, the operation comprises three steps:

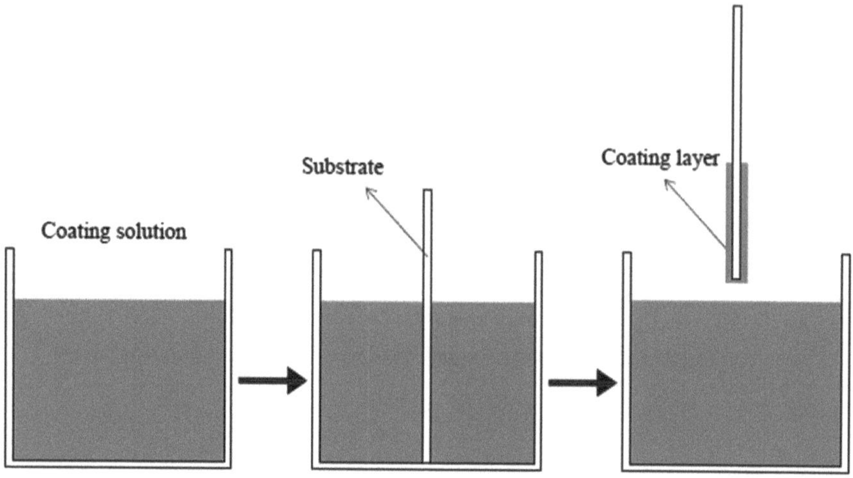

Fig. 3.6 Dip-coating process

- filling of the mold;
- compaction of dough;
- extraction of the piece ready.

Two categories of pressing are commonly used, (1) uniaxial and (2) isostatic. Both use the powder prepared by the same procedures. So, like other processes used in preparing ceramic pieces, pressing is a process that involves steps that should be carefully controlled. The pressing is a method commonly used for the preparation of inorganic membranes. Figure 3.7, in simplified form, shows the commonly used methods of pressing, isostatic and uniaxial. This method has often been used in the preparation of ceramic membranes permeable to oxygen or hydrogen. A special press machine is used to apply more than 100 MPa of pressure to press post on a compressed drive. The diameter of the disc is usually a few cm, thickness is often about 0.5 mm. Additives commonly used in pressing are the blinder, plasticizers, lubricant, and other assistants of compression. Table 3.1 presents some additives used in the technique of pressing, according to the product to be conformed. The blinder has for function to promote the lubrication and gives the piece down resistance suitable for handling, inspection and machining. Plasticizer gives more flexibility. The lubricant provides less friction between the particles and other press components. And finally, the compression AIDS promote the reduction of the friction between the particles (Reed 1995; Richerson 2006; Li 2007).

3.2.6 *Sol–Gel*

The first sol–gel process conducted was in 1845, in France, by Ebelmen in manufacture of ceramics. Generally, the sol–gel process can be understood as a route of

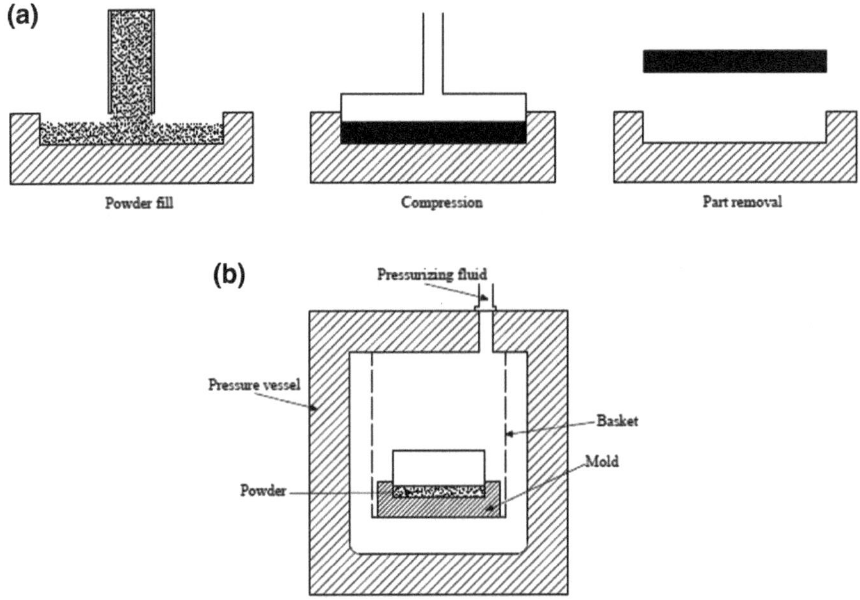

Fig. 3.7 Pressing process **a** uniaxial and **b** isostatic

Table 3.1 Additives used in the pressing method (Reed 1995)

Product	Blinder	Plasticizer	Lubrificant
Alumina	Polyvinyl alcohol[a]	Polyethylene glycol[b]	Mg Stearate
96% alumina substrate	Polyethylene glycol[c]	None	Talc[d], clay[d]
Alumina spark plug insulation	Microcrystalline wax emulsion	KOH + tannic acid	Wax, talc[d], clay[d]
MnZn ferrites	Polyvinyl alcohol[a]	Polyethylene glycol[b]	Zn stearate
Ba titanate	Polyvinyl alcohol[a]	Polyethylene glycol[b]	
Steatite	Microcrystalline wax, clay	Water	Wax, talc[d], clay[d]
Ceramic tile	Clay	Water	Talc, clay
Hotel china	Clay, polysaccharide	Water	Talc[d], clay[d]
Refractories	Ca/Na lignosulfonate	Water	Stearate

[a]Low-viscosity grade
[b]400 molecular weight
[c]20,000 molecular weight
[d]Colloidal size

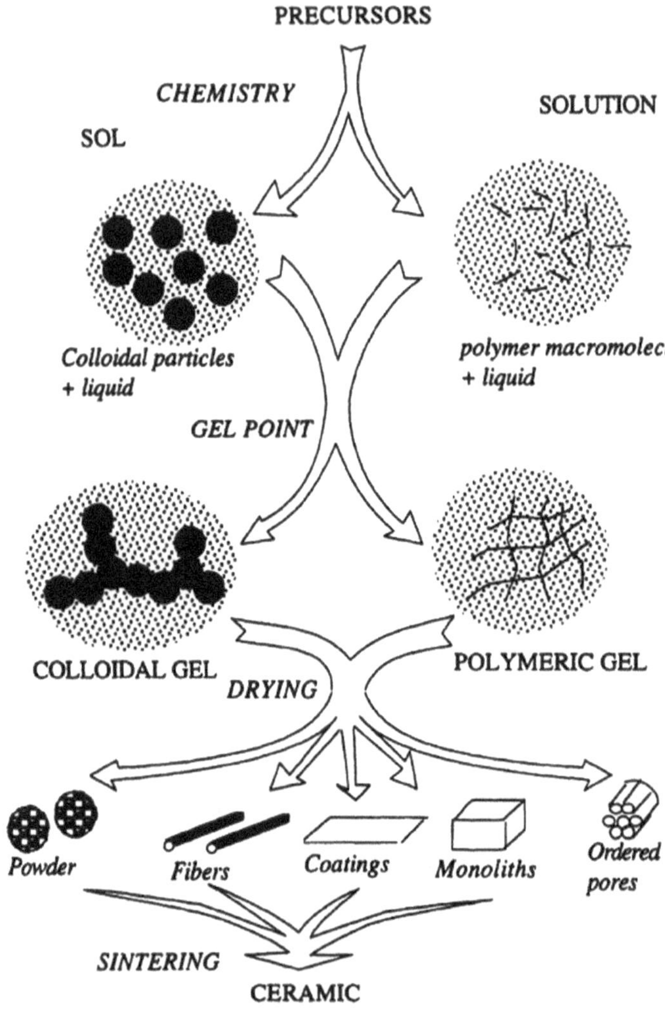

Fig. 3.8 Simplified process sol–gel (Pierre 1998)

obtaining ceramics with the intermediate stage, which includes the state of sol and/or gel. This process is not known to present a single technique, but a wide range of procedures that focuses on a single schema, which can be viewed in Fig. 3.8. This technique is not limited only to production of materials whose composition is an oxide, but also allows the production of new hybrid materials, such as organic–inorganic, that do not exist naturally. Another advantage is that the first steps of the process are carried out at low temperature, which minimizes the chemical interactions between the material and the walls of the container. In addition, the colloidal State of the solid to the liquid prevents pollution by possible dispersion of dust. And because of this, the largest implementation for this method is in the nuclear fuel industry, where pollution can be critical to the process (Burggraaf and Cot 1996; Pierre 1998).

The sol can be understood as a colloidal suspension of solid particles in a liquid, which may be a dense oxide, as an example we have Al_2O_3, SiO_2, TiO_2, ZrO_2. This type of colloid can be used to generate polymers or ceramic membranes of which particles are made. In the process sol–gel, the precursors used in the preparation of a colloid consist of a metallic element surrounded by several other ligands. An example of precursors for alumina are inorganic salts like $Al(NO_3)_3$ and organic compounds such as $Al(OC_4H_9)_3$ (Brinker and Scherer 1990). The sol–gel method was applied for the first, in the development of ultrafiltration ceramic membranes, Leenaars and Burggraaf (1985a, b, c). The main routes used in the sol–gel process are described as follows (Burggraaf and Cot 1996):

(1) Colloidal via, in which a metallic salt is mixed with water to form a sol, the sol is covered in a membrane, where it forms a colloidal gel.
(2) The path in which the polymer metal-organic precursors are mixed with organic solvent to form a sol, which is then covered in a membrane, where it forms a polymer gel.

Membranes obtained by sol–gel method have demonstrated unique properties. The most important feature of this method is the high porosity (30–70%) that the membranes obtained by this method are presented. Over the years, the method has made possible the development of membranes with pore geometry and controlled on the nanoscale. In addition, stood out by providing a good dimensional pore structure, high chemical resistance, and mechanical stability (Burggraaf and Cot 1996).

Gas separation based on molecular sieving purposes, must used ceramic membranes with pore sizes less than 1 nm. In this case, the membrane can be prepared by the route of the sol using polymeric membrane of γ-alumina prepared from colloidal sol support. It should be noted that in the course of the sol, the polymeric membrane prepared pore size is determined by the degree of branching of the polymer. A low degree of branching would result in a narrower pore system (Li 2007).

References

Amirilargani, M., Sadrzadeh, M., Sudhölter, E.J.R., Smet, L.C.P.M.: Surface modification methods of organic solvent nanofiltration membranes. Chem. Eng. J. **289**, 562–582 (2016)

Auerbach, S.M., Carrado, K.A., Dutta, P.K.: Handbook Zeolites Science and Technology. Marcel Dekker, New York (2003)

Bauser, M., Sauer, G., Siegert, K.: Extrusion. ASM International, Materials Park, Ohio (2006)

Bekkum, H.V., Flanigen, E.M., Jansen, J.C.: Introduction to Zeolite Science and Practice. Elsevier Science, The Netherlands (1991)

Blackburn, S., Lawson, T.A.: Mullite-alumina composites by extrusion. J. Am. Ceram. Soc. **75**(4), 953–957 (1992)

Boch, P., Nièpce, J.C.: Ceramic Materials: Processes. Properties and Applications. ISTE Publishing Company, Newport Beach (2007)

Brinker, C.J., Scherer, G.W.: Sol-gel science: the physics and chemistry of sol-gel processing. Academic Press, San Diego (1990)

Burggraaf, A.J., Cot, L.: Fundamentals of inorganic membrane science and technology. Elsevier Science, Netherlands (1996)

Chakraborty, S., Wang, B., Dutta, P.K.: Tolerance of polymer-zeolite composite membranes to mechanical strain. J. Membr. Sci. **518**, 192–202 (2016)

Gitis, V., Rothenberg, G.: Ceramic Membranes: New Opportunities and Practical Applications. Wiley, Weinheim (2016)

Gitzen, W.H.: Alumina as a Ceramic Material. American Ceramic Society, Ohio (1970)

Hlavác, J.: The Technology of Glass and Ceramics: An Introduction. Elsevier Science, Amsterdam (1983)

Kingery, W.D., Bowen, H.K., Uhlmann, D.R.: Introduction to Ceramics. Wiley, New York (1976)

Leenaars, A.F.M., Burggraaf, A.J.: The preparation and characterization of alumina membranes with ultra-fine pores: part 3. The permeability for pure liquids. J. Membr. Sci. **24**(3), 245–260 (1985a)

Leenaars, A.F.M., Burggraaf, A.J.: The preparation and characterization of alumina membranes with ultra-fine pores: part 4. Ultrafiltration and hyperfiltration experiments. J. Membr. Sci. **24** (3), 261–270 (1985b)

Leenaars, A.F.M., Burggraaf, A.J.: The preparation and characterization of alumina membranes with ultrafine pores: part 2. The formation of supported membranes. J. Colloid Interface Sci. **105**(1), 27–40 (1985c)

Li, K.: Ceramic Membranes for Separation and Reaction. Wiley, London (2007)

Li, P., Wang, Z., Qiao, Z., Liu, Y., Cao, X., Li, W., Wang, J., Wang, S.: Recent developments in membranes for efficient hydrogen purification. J. Membr. Sci. **495**, 130–168 (2015)

Onoda, G.Y., Hench, L.L.: Ceramic Processing Before Firing. Wiley, New York (1978)

Peyki, A., Rahimpour, A., Jahanshahi, M.: Preparation and characterization of thin film composite reverse osmosis membranes incorporated with hydrophilic SiO_2 nanoparticles. Desalination **368**, 152–158 (2015)

Pierre, A.: Introduction to Sol–Gel Processing. Springer, New York (1998)

Pijolat, M., Dauzat, M., Soustelle, M.: Influence of additives and water vapour on the transformation of transition aluminas into alpha alumina. Thermochim. Acta **122**, 71–77 (1987)

Rauwendaal, C.: Understanding Extrusion. Hanser, Ohio (1998)

Reed, J.S.: Principles of Ceramics Processing. Wiley-Interscience, New York (1995)

Rice, R.W.: Ceramic Fabrication Technology. Marcel Dekker Inc, New York (2003)

Richerson, D.: Modern Ceramic Engineering: Properties, Processing, and Use in Design. Taylor & Francis Group, Boca Raton (2006)

Riedel, R.: Handbook of Ceramic Hard Materials. Wiley-VCH, Weinheim (2000)

Santos, P.S., Santos, H.S., Toledo, S.P.: Standard transition aluminas. Electron Microscopy Studies. Mater. Res. **3**(4), 104–114 (2000)

Somiya, S.: Handbook of Advanced Ceramics. Elsevier Science, London (2003)

Szostak, R.: Molecular Sieves Principles of Synthesis and Identification. Springer, London (1989)

Tsuru, T., Narita, M., Shinagawa, R., Yoshioka, T.: Nanoporous titania membranes for permeation and filtration of organic solutions. Desalination **233**, 1–9 (2008)

Tsuru, T., Sudoh, T., Yoshioka, T., Asaeda, M.: Nanofiltration in non-aqueous solutions by porous silica–zirconia membranes. J. Membr. Sci. **185**, 253–261 (2001)

Tsuru, T., Sudou, T., Kawahara, S., Yoshioka, T., Asaeda, M.: Permeation of liquids through inorganic nanofiltration membranes. J. Colloid Interface Sci. **228**, 292–296 (2000)

Wang, B., Sun, C., Li, Y., Zhao, L., Ho, W.S.W., Dutta, P.K.: Rapid synthesis of faujasite/polyethersulfone composite membrane and application for CO_2/N_2 separation. Microporous Mesoporous Mater. **208**, 72–82 (2015)

Xu, M., Wang, S.L., Shen, F., Ji, J.Y., Dai, W.B.: Study on the influential factors for preparing transition alumina whiskers. J. Alloys Compd. **695**, 2865–2869 (2017)

Chapter 4
Ceramic Membranes Characterization

This chapter is about some of the main methods used in the ceramic membranes characterization. Some of these methods are permeability, solute rejection assays, porosimetry by mercury intrusion, scanning electron microscopy, X-ray diffraction, among others. The choice of methods to be used will depend on the need of knowing which characteristic of the material are in fact relevant. When knowing the material's characteristics, it will be possible to verify where and when it will be applied, and under which conditions. Usually, the preparation of a new material will demand a greater amount of analysis, if the material has not been studied or if there are no reachable characteristics available on the literature, a more detailed investigation will be required.

4.1 Permeability

The membranes can be characterized on permeability when applied to different operation conditions, as, for example, under different feed current pressures or temperatures. The permeability is, by all means, the quantity of mass or volume fluid which goes through the membrane. This way of characterizing is one of the most simple and significant tests of determining some important characteristics of the membrane, and can be used to determine the size of the membrane pores. In one of the bench tests, the membrane is allocated in modules and its permeate flow is measured over the time. These measures show the membrane's behavior through a determined time. When performing tests with a new membrane, it is possible to get information on the size and pores distribution, and on the pores conductivity, in other words, if the pores allow the fluid passage through the membrane. Another view to be done, based on permeability, is the quality of the flow over time. It means, when observing a decrease on the permeated flow along the permeation, this effect can be attributed to a possible fouling presented by the membrane. This phenomenon indicates the need for a chemical or physical membrane cleaning, and,

D. da Silva Biron et al., *Ceramic Membranes Applied in Separation Processes*,
Topics in Mining, Metallurgy and Materials Engineering,
DOI 10.1007/978-3-319-58604-5_4

if the cleaning process is not convincing, it will be necessary to discard the membrane, because its lifetime might have ended (Ho and Sirkar 1992; Mallada and Menéndez 2008).

The pores size of a membrane can be determined by Hagen–Poiseuille's equation (Eq. 4.1) considering that its pores are cylindrical. Depending on the shape of its pores, Konezy–Carman's equation (Eq. 4.2) can be used which considers that the pores present the shape of interstices between packaged spheres.

Hagen–Poiseuille's Equation

$$J = \frac{\varepsilon \cdot rp^2}{8 \cdot \mu \cdot \tau} \cdot \frac{\Delta P}{\Delta x} \tag{4.1}$$

$$rp = \sqrt{4\frac{8 \cdot \mu \cdot \tau \cdot \Delta x}{n\,\pi} \cdot \frac{J}{\Delta P}} \tag{4.2}$$

$$P = \frac{\varepsilon \cdot rp^2}{8 \cdot \mu \cdot \tau \cdot \Delta x} \tag{4.3}$$

where K is the permeated flow ($m^3\ m^{-2}\ s^{-1}$), $\varepsilon = n\ \pi\ r^2$ is the membrane's porosity, ΔP is the transmembrane pressure, μ is the fluid's absolute viscosity (Pa s), rp is membrane pore radius (m), τ is the pore tortuosity, Δx the membrane thickness or pore length (m), n is the amount of pores presented by the membrane, and P is hydraulic permeability (m $Pa^{-1}\ s^{-1}$).

Equation 4.4 allows the calculation of the membrane pore radius. Assuming that the pores are cylindrical, parallel to each other, identical, perpendicular to the membrane area and that the fluid crosses the membrane from the feed current to the permeated with no restrictions, it means that the tortuosity is equal to 1. Finally, with (Eq. 4.5) it is possible to obtain the hydraulic permeability of the membrane.

Carman–Kozeny's Equation

$$J = \frac{\varepsilon^3}{CK \cdot \mu \cdot A^2 \cdot (1 - \varepsilon)} \cdot \frac{\Delta P}{\Delta x} \tag{4.4}$$

$$P = \frac{\varepsilon^3}{K \cdot \mu \cdot A^2 \cdot (1 - \varepsilon)^3 \cdot \Delta x} \tag{4.5}$$

where CK is Carman–Konezy's constant, A is the membrane's specific superficial area. CK's constant depends on the pore shape and membrane's tortuosity.

A way of experimentally determining the hydraulic permeability of the membrane is to express the medium flow values, related to different applied pressure, as

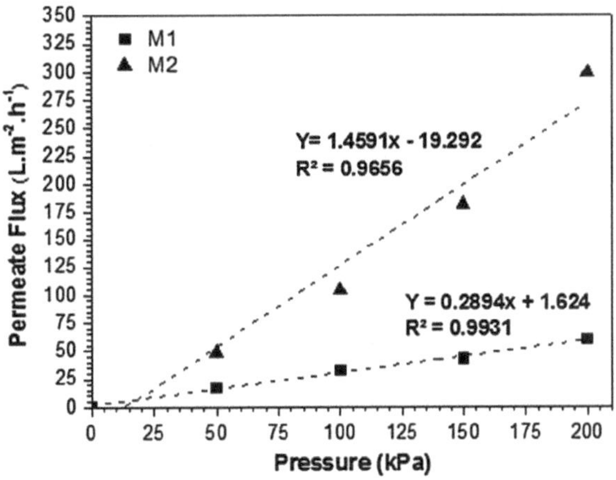

Fig. 4.1 Behavior of the medium permeate flow, with distilled water, in relation to the different transmembrane pressures applied. M1: α-alumina membrane with average pore size of 0.4 μm and M2: mullite membrane with average pore size of 0.6 μm

can be visualized on Fig. 4.1. This technique is widely used by many authors (Porter 1990).

As can be observed in Fig. 4.1, the hydraulic permeability will depend on the characteristics of each membrane, as pore tortuosity, porosity, amount of effective pores, among others (Damas et al. 2012). In the example above, it is verified that the permeability corresponds to the coefficient of the line. To the α-alumina membrane, the corresponding value is 28.94 L m^{-2} h^{-1} bar^{-1} and to the mullite membrane it corresponds to 145.91 L m^{-2} h^{-1} bar^{-1}.

4.2 Solute Rejection Assays

The solute rejection assay is another widespread technique among researchers, which have as goal the molecular weight cut-off from solutes. This analysis is used by manufacturers, in the characterization of ultrafiltration membranes. In a simple context, this molecular weight cut-off (MWCO) is the rejection of the membrane (around 90%) to a solute of known molar mass (Calvo et al. 2011; Zeidler et al. 2014; Chen et al. 2017). Based on the solute rejection values of different molar masses, it is possible to obtain the retention curve of a certain membrane, Fig. 4.2. The advantage of such technique is the achievement of a direct and indisputable result, with no assumptions or additional calculus. Still, using other tests, it is possible to obtain data on the membrane pores size and relate them among the rejection levels. The second advantage is that this kind of analysis can be applied at any moment of the membrane operation and it can detect only the active pores. Some other tests, such as mercury porosimetry, require a tests' stop and the removal of the module membrane. Finally, the costs with these kinds of characterization are way lower when compared to other techniques (Li 2007).

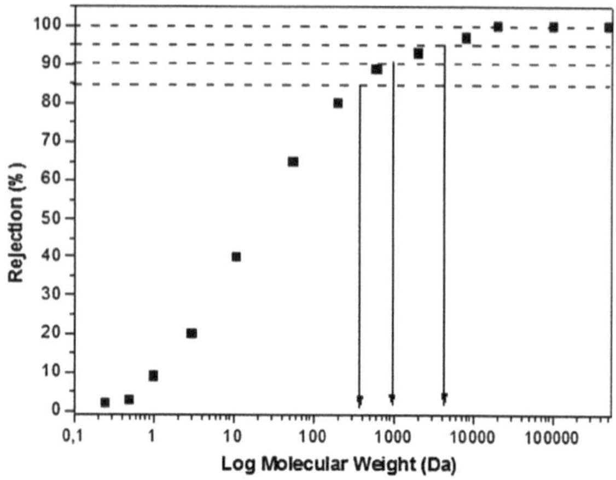

Fig. 4.2 Example of retention curve of a membrane any. In this case, the MWCO of the membrane is around 400 Da to a 85% rejection, 1000 Da to a 90% rejection and 4000 Da to a 95% rejection

Despite the advantages of this technique, it does not allow a complete characterization that could relate the molar mass of the molecules or its dimensions to the pore dimensions or the membrane's porosity. Still, one must consider the molecule characteristics and its interaction to the membrane material. An example of interaction that occurs, most of the time, is the polarization by amount, which must be considered. The increase in solute molecule amount can alter its dimensions and affect the membrane absorption. Consequently, differences in the retention curve occur, depending on the process parameters (Calvo et al. 2011).

During the membrane retention tests, the solute is added to the feed flow and monitored on the permeate outlet. In this test, the only variant is the solute itself and its amount. The choice of the solute depends on the tested membrane, the required results, tests frequency, and the solute's availability and sensitivity. A great variety of solutes can be used in this test, as, for example proteins, polymers and ions to liquid fluids and gases such as oxygen and nitrogen to gaseous fluids (Arthanareeswaran et al. 2008). Table 4.1 presents some of the examples of proteins used on the MWCO determination.

Table 4.1 Proteins used on the MWCO determination and its physical characteristics (Adapted from Arthanareeswaran et al. 2008)

Protein	Radius (Å)	Molecular weight (Da)
Cytochrome c	15.5	12,400
Myoglobin	17.5	17,000
Trypsin	21.5	20,000
Pepsin	28.5	35,000
Egg albumin	33.0	45,000
Hemoglobin	27.5	64,000
Human serum albumin	30.3	67,000
Bovine serum albumin	45.0	69,000
IgG-Immunoglobulin	65.0	150,000

In general, the retention tests are performed with PEG or Dextran solutes with different molecular masses or high molecular weight proteins, as trypsin, pepsin, egg albumin, and bovine serum albumin. And, to calculate the rejection of these solutes presented by the membrane, (Eq. 4.6) is used.

$$R(\%) = \left(1 - \frac{Pc}{Fc}\right) \cdot 100 \tag{4.6}$$

where R(%) is the percentage rejection, Pc is the amount of the solute in the permeated, and FC is the amount of solute in feed.

The usage of PEGs mixtures tends to underestimate the MWCO and the membrane pore size, as the solutes of greater mass block the permeation of the ones with lower mass. Shang and contributors indicate the influent assessment of PEGs amount and the transmembrane pressure variety on MWCO dimension. Generally, an analysis of PEGs amount is done by high-performance liquid chromatography (HPLC) (Shang et al. 2017). PEG's usage, as solute, on the MWCO analysis allows us to determine the pore diameter on (Eq. 4.7) (Bruggen and Vandecasteele 2002), which relates the molecular mass of the solute. However, some claims are to be considered: the membrane pore size follows a regular distribution, the separation mechanism is based on the exclusion of size and the diffusion of the solute is insignificant.

$$dp = 0.065 \cdot (MWCO)^{0.438} \tag{4.7}$$

According to Biron et al. 2015 the MWCO analysis with the proteins allows to measure the pore radium, the porosity and amount of pores presented by the membranes. The solute radium can be determined by (Eq. 4.8), which considers the rejection result, in percentage, established by the MWCO. Having the result of the pore size, it is possible to calculate the porosity and amount of membrane pores, Eqs. 4.9 and 4.10, respectively (Arthanareeswaran et al. 2009; Arthanareeswaran and Thanikaivelan 2010; Biron et al. 2015).

$$R = 100 \cdot \left(\frac{r}{\%PR}\right) \tag{4.8}$$

$$\varepsilon = \frac{3\pi \cdot \mu \cdot J}{R \cdot \Delta P} \tag{4.9}$$

$$n = \frac{\varepsilon}{\pi \cdot R^2} \tag{4.10}$$

where R is the pore radium presented by the membrane, r is the solute medium radium, %RP is the protein rejection by the membrane (90%) where MWCO occurs, ε is the membrane porosity, μ the absolute viscosity of the solvent, often using distilled water, J is the permeated flow, ΔP is the transmembrane pressure applied, and n the amount of pores.

4.3 Porosimetry by Mercury Intrusion

The beginning of porosimetry by mercury intrusion's history took place in around 1842, when J. Smetana mentioned the possibility of forcing the entrance of mercury in wood, to prove its porous structure. Even though Washburn, in 1921, mentioned a technique in the determination on pore size distribution, using applied pressure data and mercury volume, it was Henderson and contributors who determined the pores distribution of the first materials. Henderson and others, in 1940, verified clay's and bauxite's porous structure, based on Washburns suggestions, to a 6 MPa pressure. Based on previous authors, Ritter and Drake, in 1945, published the first papers related to this technique. Ritter and Drake not only publicized a vast amount of experimental data, but also described the construction and operation of the mercury porosimetry equipment. Such discoveries were made the basis for many upcoming papers, and, until today, this technique is considered one of the most important techniques used on porous materials (Brakel 1981).

The porosimetry by mercury intrusion technique consists in placing a porous sample in a closed recipient, then, removing the gas and insert mercury in the sample pores. To the mercury insertion occur, it is necessary to add pressure gradually. This way, for each increase of pressure, the mercury volume added inside the pores is measured. The variations of mercury volume are detected under the shape of conductance variation, being done the calibration of the port sample, named penetrometer. The choice of mercury as liquid of mediation to the porosity variation is due to its non-wetting characteristic when in contact with most solid substances. This characteristic makes mercury a substance which does not penetrate naturally the pores, making it necessary to apply a pressure that obligates its penetration (Masuda et al. 2006; Gitis and Rothenberg 2008).

Basically, this technique presents two steps. On the first step, all the gaseous fluid is removed from the sample pores and then the sample is soaked in mercury. Gradually, pressure increases and mercury is inserted to the porous environment. During the process, mercury volume is discriminated according to the applied pressure. This step is named intrusion. On the second step, there is a relief in the pressure down to atmospheric pressure, and, consequently, an outlet of part of the mercury in the pore sample happens. This process, called extrusion, completes the technique cycle. As a result, we get what is called hysteresis, which is a systematic deviation between the intrusion and extrusion curves, coming from the amount of mercury still on the material. Finally, the samples contaminated with mercury must be discarded in a safe way (Brakel 1981; Masuda et al. 2006).

Figure 4.3 presents an example of the porosimetry by mercury intrusion technique. The data refer to an α-alumina porous membrane. The figure presents the cumulative volume of mercury, per unit of sample mass, due to applied pressure.

The porosimetry by mercury intrusion technique is often used for porous samples, where the pore diameter varies from 0.003 to 1000 μm. From the experimental data obtained by the porosimetry by mercury intrusion, it is possible to determine the diverse porous sample characterization parameters, such as, pore size

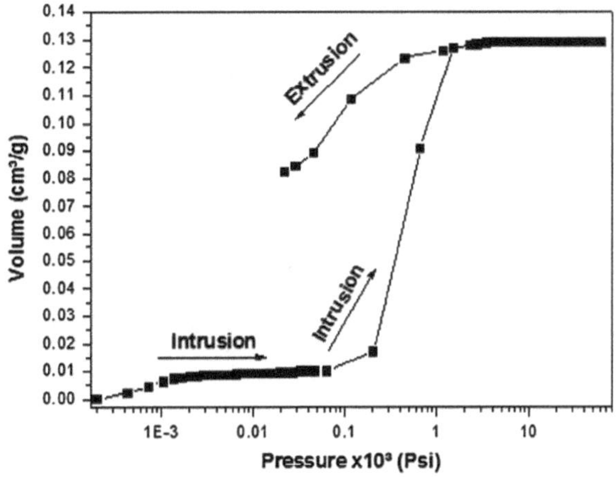

Fig. 4.3 Experimental result of porosimetry by mercury intrusion of an α-alumina membrane

distribution, average pore size, total pore volume, superficial area, among others. This technique allows the characterization of a wide variety of materials, such as porous ceramic materials, porous polymeric materials, refractory materials, catalysts, rocks, resins, ceramic and polymeric membranes and others (Brakel 1981).

As previously mentioned, the hysteresis is understood as a deviation between the extrusion and intrusion. Because of that, there must be an analysis on the intrusion of mercury data (Masuda et al. 2006). Figure 4.4 presents two of the main ways to analyse the pore size distribution, based on mercury intrusion data.

As can be seen on Fig. 4.4a and b it is possible to extract data on the medium pore (0.32 μm) and its variation (0.04 and 3.4 μm) and the pore volume (0.12 cm^3/g). Having these values, it is still possible to determine, indirectly, the membrane porosity, its superficial area and other important data (Gitis and Rothenberg 2008). Some of the equations of the main parameters that can be obtained from porosimetry by mercury intrusion will be presented. Currently, with more sophisticated equipment, it is high likely that the report the equipment gives already provides such data.

- Specific apparent mass
  Mercury cannot penetrate pores with diameters lower than 1000 μm on low pressure (0.01 atm). Consequently, the volume occupied by the porous solid, Vs, is equal to the sum of the volume of the solid with the pore volume's diameter lower to 1000 μm. This way, the specific apparent mass, ρ_a, can be calculated by (Eq. 4.11).

$$\rho_a = \frac{m_s}{Vs} \tag{4.11}$$

where m_s is the solid mass.

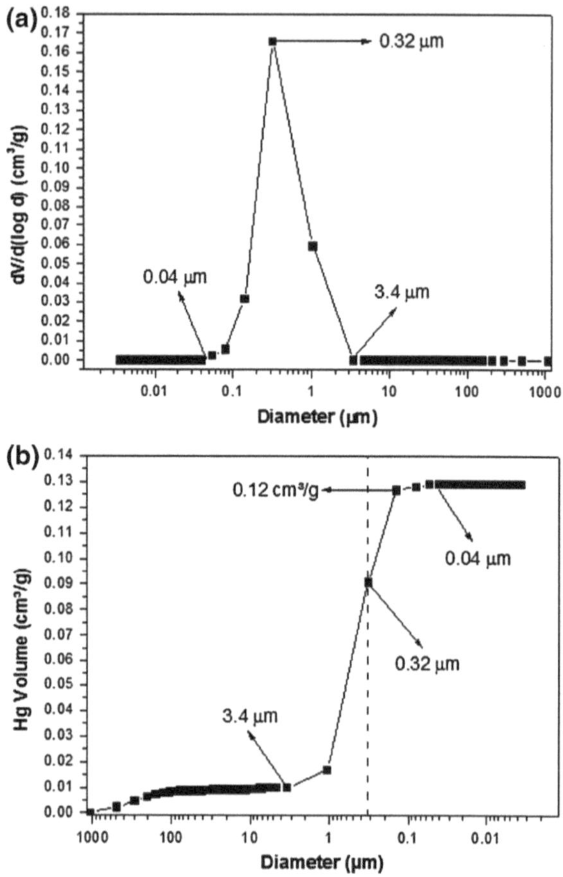

Fig. 4.4 Experimental result of the two main ways to present the obtained data on the porosimetry by mercury intrusion technique of an α-alumina membrane. **a** Derivative profile of cumulative curves, function dV/(log d), and **b** intruded mercury volume

- Specific real mass
  When considering that under the maximum pressure (4000 atm) all solid pores are filled by mercury, the volume occupied on this pressure, Vm, is equal to the volume of the solid sample matrix. This way, the specific real mass, ρ_r, is calculated by (Eq. 4.12).

$$\rho_r = \frac{m_s}{Vm} \tag{4.12}$$

- Pores volume
  Equation 4.13 allows us to calculate the pores volume on the analyzed material.

$$Vp = Vs - Vm \tag{4.13}$$

where Vp is equal to the pores volume.

- Sample porosity
  Porosity can be obtained, when having the specific apparent and real mass values, by (Eq. 4.14).

$$\varepsilon = \frac{\mathrm{Vp}}{\mathrm{Vs}} = 1 - \frac{\rho_a}{\rho_r} = 1 - \frac{\mathrm{Vm}}{\mathrm{Vs}} \tag{4.14}$$

- Superficial area
  When assuming that the pores of a specific sample are cylindrical, it is possible to calculate the specific area value by mass unit.

$$A_n = \frac{4 \cdot (V_n - V_{n-1})}{D_n} \quad n = 1, 2, 3, \ldots, N \tag{4.15}$$

$$D_n = \frac{D_n + D_{n-1}}{2} \quad n = 1, 2, 3, \ldots, N \tag{4.16}$$

$$A = \sum_{n=1}^{N} A_n \tag{4.17}$$

where A is the superficial area, V the pore volume per sample gram and D the pore sample diameter.

With (Eq. 4.15), we can calculate the incremental values of the superficial area, to a sum of N intervals. The medium pore diameter in this interval is calculated by (Eq. 4.16). Finally, the total superficial area is obtained by (Eq. 4.17), adding up all the incremental areas.

4.4 Gas Adsorption and Desorption—BET

The gas adsorption and desorption technique, used in the determination of parameters of a porous material, uses the concepts of Brunauer–Emmett–Teller (BET). This method, just as for porosimetry by mercury intrusion, allows the determination of pore size and membrane superficial area. The method consists in exposing a porous material to a flow of gaseous molecules, which adsorb the pores. The superficial adsorption can occur in two different ways: physisorption and chemisorption. These mechanisms are distinct by the nature of intermolecular attractions between the molecule and the surface—being the first a physical interaction and the second an interaction which involves chemical bonds (Gitis and Rothenberg 2008; Tarleton 2015). Table 4.2 presents the characteristics of both kinds of adsorption.

Table 4.2 Characteristics of physical and chemical adsorption (Tarleton 2015)

Physisorption process	Chemisorption process
Low adsorption energy, 1–2 kcal/mol;	High adsorption energy, 10–100 kcal/mol;
The interaction forces are Van der Waal's forces;	The attraction forces are chemical bonds;
Usually, interaction occurs in low temperature and decreases with temperature increase;	This interaction occurs with high temperature;
It is reversible;	It is irreversible;
It is related to the ease of gas liquefaction;	Usually, the adsorption extension is not related to gas liquefaction;
It is not a process that requires specificity;	It is a process that demands high specificity;
It forms multimolecular layers;	It forms monomolecular layers;
It does not require activation energy	It requires activation energy.

Physical and chemical adsorption can occur, however, physical adsorption is the most important, because it is applied to all surfaces and gases (Gitis and Rothenberg 2008). Due to the low energy required to physical adsorption, to increase the quantity of gas adsorbed to the material, the experiments are made on low temperature (77 K to the nitrogen). The commonly used gases are nitrogen, argon or carbon dioxide, being nitrogen the most utilized. The quantity of gas that will adsorb to the material will depend on its porosity or on the applied pressure. Initially, the experiments done using this technique, worked on low pressure (1 atm), what allows analyzing pores with higher dimensions. As pressure increases, the gas will adsorb on lower pores. To every increased pressure, the gas volume which is adsorbed to the material is measured (Li 2007).

The gas adsorption isotherms when under temperatures reaching the gas dew point show two regions to most of adsorbents: low isotherm pressures are concave, the higher ones are convex, both related to the pressure axis (Brunauer et al. 1938). To better understanding, Fig. 4.5 presents five of the main adsorption isotherms, which include Brunauer, Deming, Deming, and Teller's adsorption isotherms (BDDT).

Type I isotherm is typical for microporous solids and chemisorption isotherms. The one named Langmuir isotherm, Fig. 4.5a, presents a fast increase of nitrogen adsorbed until a maximum value, which is maintained as steam pressure increases. The maximum value adsorbed represents the final stage of monolayer adsorption. Type II isotherm is observed in nonporous solids and refers to physical adsorption of multiple reversible layers, presenting an inflection point which is attributed to the monolayer formation. After the monolayer adsorption, the additional increase of partial pressure results on extensive adsorption and on multilayers addition. Type III and V isotherms are typical for steam adsorption. Type III isotherm is characterized by its convexity on the partial pressure axis. Type V isotherm, Fig. 4.5e, similar to type III isotherm, is commonly observed in homogeneous flat

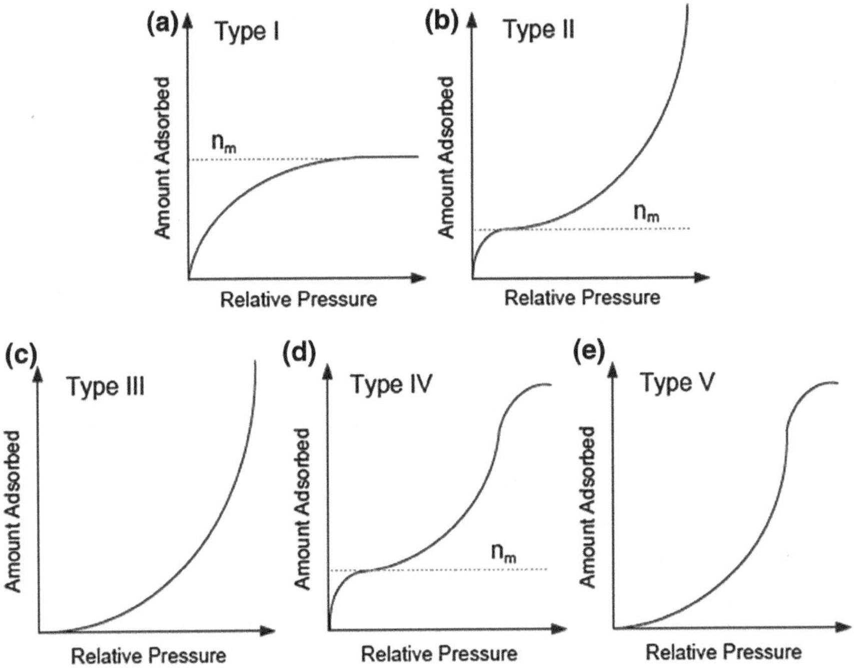

Fig. 4.5 Representation of adsorption curves, types I, II, III, IV and V of BDDT (adapted from Tarleton 2015)

adsorbents. In this case, the adsorbate interacts preferably with the monolayer than with the adsorbing surface, due to lower adsorption energy compared to liquefaction energy. Still, type III and V isotherms show very low interaction on systems that contain macropores or mesopores. Finally, type IV isotherm, Fig. 4.5d, is similar to type II isotherm, having inflection as monolayer formation. This isothermal is popularly shown by many mesoporous industrial adsorbents and xerogels of inorganic oxides, which monolayer surface coverage of pore walls is followed by capillary condensation or pore filling. Type IV and V isotherms have a hysteresis loop; the desorption on mesopores occurs under lower pressures than in those which pore adsorption would occur on similar size. Due to capillary condensation, the membranes with mesopores show unique adsorption/desorption behavior of types IV and V (Li 2007; Tarleton 2015).

4.5 Scanning Electron Microscopy (SEM)

Scanning electron microscopy (SEM) development occurred parallelly to transmission electronic microscopy (TEM). SEM allows a surface view of bulky samples. Scanning electron microscopes were widely accepted due to their wide range of application and its ease on sample preparation. Due to their wide application on morphological microstructure of ceramic membranes characterization, this method became indispensable on analyzing ceramic membranes (Echlin 2009).

Generally, SEM has these following main attributes (Goldstein et al. 2003):

- Obtainment of smooth or rough surfaces images, with field depth and high resolution;
- In a great majority of materials, the images generated are of easy understanding, with three-dimensional appearance;
- Possibility of process of digital signs, manipulating and processing images;
- With other equipment support (as EDS), it is possible to perform elemental microanalysis.

Such technique has its main application on visualizing irregular surfaces, being able to identify a characteristic topography. TEM technique allows the obtainment of direct images, in which rays go through the sample, passing through lenses and form a real image. However SEM technique generates indirect images, meaning there is no optical path between the sample and image, having generation and visualization systems separated from the image. The obtainment of a low magnification images (<1000 X) is relatively simple with SEM equipment. However, to obtain the needed information with this technique, it is necessary to know microscopy's main functions and electron beam parameters that affect it (Goldstein et al. 2003; Echlin 2009).

The two main components of SEM is electronic column and console control. Electronic column consists in an electron cannon and two or more electronic lenses, which influenced the electron's paths in an evacuated tube. Electronic column basis is composed of a vacuum pump which allows a vacuum of about 10^{-4} Pa. Console control consists of a visual display of cathode rays tube (CRT), buttons, computer keyboard which allows electron beam control (Goldstein et al. 2003).

SEM's principal consists in using electron beam of small diameter to explore the surface of a small sample of ceramic membrane. The sample scanning occurs by successive lines and the detector's signal is transmitted to a cathode display whose scanning is perfectly synchronized with the one of incident beam. The beam can be guided through deflection coils system, to scan the sample surface. The signal collected by the detector is utilized to modulate the screen's brightness, allowing observation. Most pieces of SEM equipment use as electron resource, a heated tungsten filament (W). The beam is accelerated by high tension created between the filament and the anode. Then, it is focused by the sample over a series of three electromagnetic lenses with a spot lower than 4 nm. The beam interacting with the sample produces electrons and photons which can be collected by accurate detectors and converted into a video signal. The electrons have much lower wavelengths than the photons of visible light, and, therefore, give more detailed micrographs. An electronic microscope provides between 10,000 and 500,000 times of magnification (Goldstein et al. 2003; Echlin 2009).

When analyzing a ceramic membrane sample, we verify that electrons interact with membrane atoms, producing signals of surface response. These signals include the secondary electrons. SEM secondary electrons result from the interaction of electronic beam with the sample material. These resulting electrons are of low voltage <50 eV, and form high-resolution images. On the physical configuration of SEM instruments, only secondary electrons produced near the surface can be detected. The image contrast is given, overall, the sample's relief, which is the main way of forming image. The secondary electrons generated by interactions between electron and atom of the sample have a free path of about 2–20 nm and that is why, only those generated onto the surface can be reemitted (Goldstein et al. 2003).

Figure 4.6 presents an image of a ceramic membrane of α-alumina obtained by scanning electronic microscopy. Many pieces of SEM equipment have, linked, an

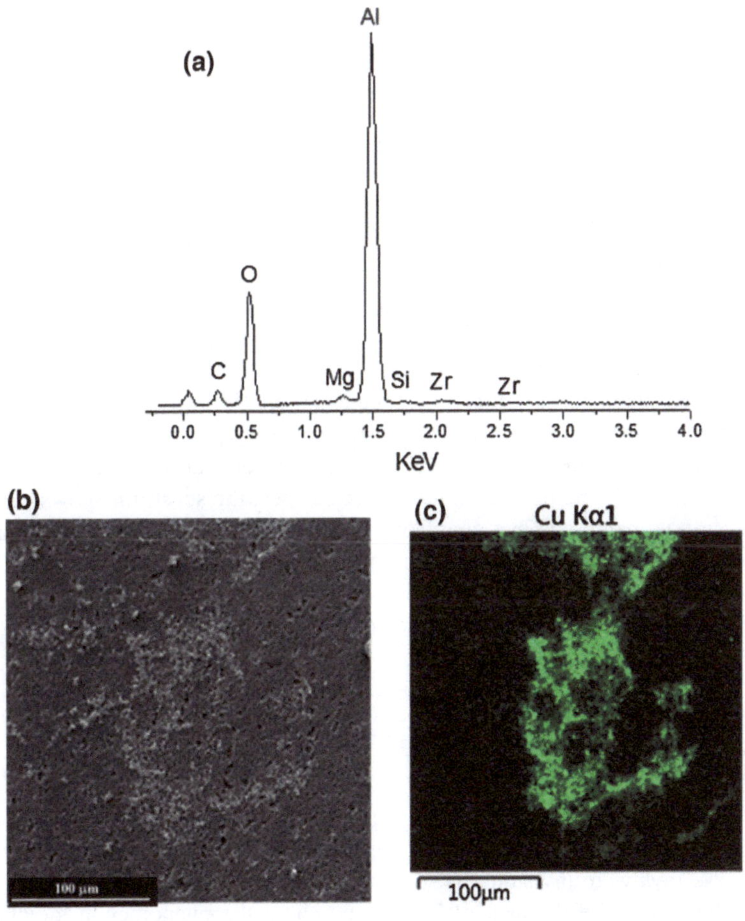

Fig. 4.6 Example of result of an analyses of scanning electronic microscopy of a ceramic membrane of α-alumina. **a** EDS spectrum, **b** SEM and **c** chemical composition map obtained by EDS, which shows copper particles on the membrane's surface

X-ray spectroscopy equipment for energy dispersion, which allows elementary characterization of the ceramic membrane. On the following chapter, this equipment will be given emphasis.

4.6 Energy-Dispersive X-Ray Spectroscopy (EDS)

By and large, the energy-dispersive X-ray spectroscopy (EDS) technique is a very common technique used for chemical characterization of ceramic membranes, allowing the construction of specters and maps of elemental chemical composition. Even though it is a spectroscopy technique, many EDS's pieces of equipment come linked to SEM equipment. It is one of the variants of X-ray fluorescence spectroscopy (Goldstein et al. 2003).

During the analysis, a semiconductor material detector is used to classify X radiation according to the energy, instead of the wavelength. The detector is a silicon crystal doped with lithium, which is polarized by electrodes on both surfaces. The photoelectron scattering on the silicon creates free electron-hole pairs on the semiconductor band structure which are separated by polarization applied through the detector, being the collected load on the electrodes' surface. The detector's function is to convert the X-ray energy in current pulses. These impulses are amplified, digitalized, and fed in a multichannel analyser or in the computer's memory which stores them in the correct place. The voltage pulses are transferred to a multichannel analyser, which has the order of 1000 channels, each one corresponding to one voltage range. When a voltage pulse hits the detector, it is allocated to the appropriate channel and its value and analyser stores all the specter, which can be obtained in seconds or minutes (Watt 1997).

Figure 4.6 exemplifies the application of the technique when used in the characterization of ceramic membranes. Figure 4.6a presents an EDS spectrum related to a ceramic membrane of α-alumina. As we can see, the spectrum shows elements on the ceramic membrane in a qualitative way, which does not allow us to evaluate in a quantitative way each element. On the other hand, Fig. 4.6c shows us a figure of an α-alumina ceramic membrane's chemical composition containing in its structure elemental copper, also generated by EDS technique.

4.7 X-Ray Diffraction (XRD)

The X-ray was discovered in 1895 by the German physics Röntgen and was called this way because its nature was unknown that historical moment. Differently from light, these rays were invisible, but they affected the photographical film the same way as light. Despite the lack of knowledge about the phenomenon, quickly the X-ray were utilized by doctors on the visualization of broken bones, and, later by engineers that wanted to see the internal structure of materials (Cullity and Stock

2014). In 1911, von Laue, Friederich e Knippin were the first to experimentally study of characterization of materials by X-ray diffraction. On their work, the first X-ray diffraction patterns were obtained, related to the rock salt. Later, in 1917, Bragg, Von Laue, Darwin and Ewald established the theoretical bases of the technique (Browen and Tanner 2005). From these papers, the technique gained prominence and until today is one of the main techniques utilized in crystalline material's characterization.

The X-ray diffraction phenomenon occurs when a wave hits a row of obstacles regularly separated. When this occurs, the obstacles are able to disperse the wave and have comparable spacing to the wavelength. The X-ray can hit the material and be elastically spread, with no loss of energy by the atom's electrons. Besides, diffraction is a consequence of relation of specific established phases between one or more waves that were spread over the obstacles. The crystalline material atoms make the beam diffract and, using the right angle measuring equipment with the right intensity of such diffracted beams, it is possible what is called diffractogram (Pecharsky and Zavalij 2005).

Figure 4.7 presents where the X-ray is found in the electromagnetic radiation spectrum. As we can see, the X-ray are an electromagnetic radiation with high energy (10^2 and 10^6 eV) form and short wavelengths (0.001–10 nm).

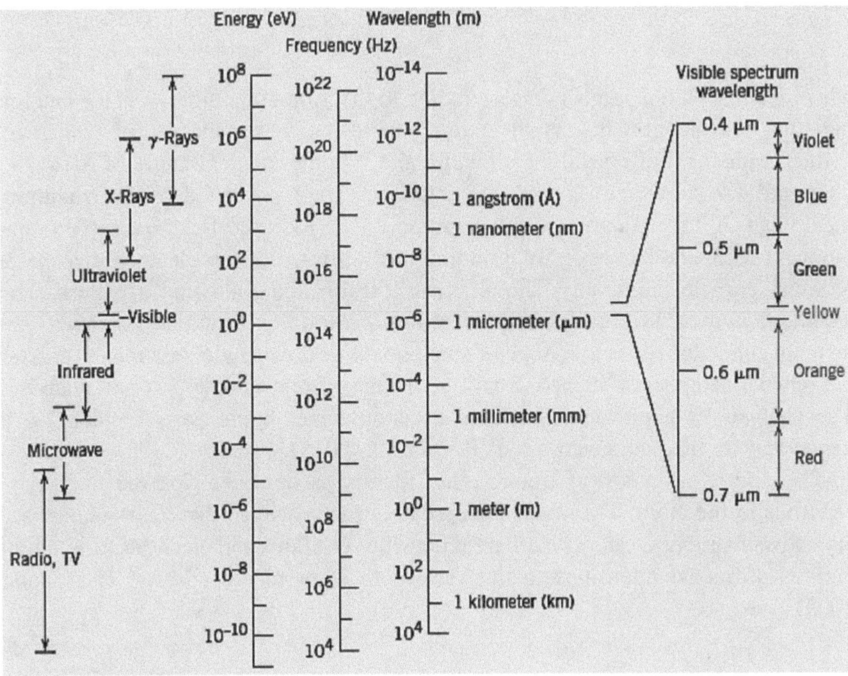

Fig. 4.7 Electromagnetic radiation spectrum. Including energy levels and wavelengths (Adapted from Callister and Rethwich 2009)

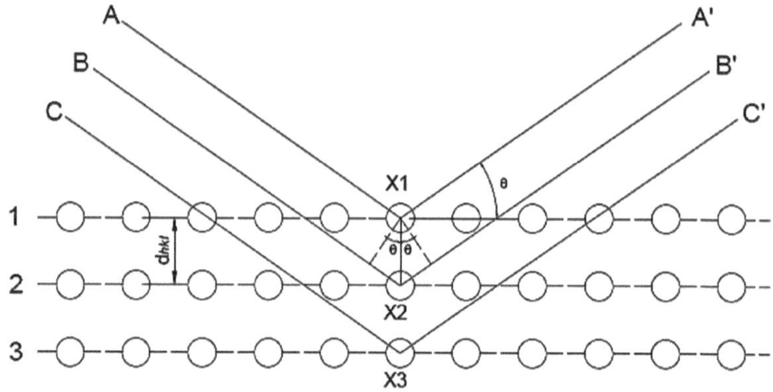

Fig. 4.8 X-ray Diffraction by planes of atoms

Considering two or more parallel atomic planes (1, 2 and 3) of a crystalline structure, Fig. 4.8, which have the same Miller indices (*h k l*) and are separated by interplanar spacing (d_{hkl}), the conditions to occur the X-rays diffraction, constructively, will depend on the difference travelled by X-rays and the wavelength of the incident radiation. Such condition is made clear on Bragg's Law, (Eq. 4.18) (Callister and Rethwich 2009).

$$n\lambda = 2 \cdot d_{hkl} \cdot \text{sen}\,\theta \tag{4.18}$$

where n is the whole number corresponding to the diffraction order, λ is the incident radiation wavelength, d_{hkl} is the interplanar distance for the set of planes *hkl* (Miller's indexes) of crystalline structure and θ is the incident angle of X-ray.

Figure 4.9 shows an example of a diffractogram presented by an α-alumina membrane. In this example, a pulverization method is used for the sample preparation. The method consists in grinding and sifting the sample to a size of an ultrafine particle, preferably, shorter than 150 μm and following a pattern. This technique is more utilized in the preparation of ceramic membrane's samples, and its main characteristic is the usage of small quantities of sample, less than 1 g. After the powder sample is obtained, it will be displayed in a sample port and analyzed. The analysis of a powder sample of a ceramic membrane, lasts around 2–3 h, depending on the parameters used (Biron et al. 2015).

The diffracted intensity, among other factors, is dependent on the number of electrons in the atom. The atoms are spread in the space, so that many planes of a crystallised structure can have different densities of atoms and electrons, originating various diffracted intensities to the various existing planes (Bowen and Tanner 2005).

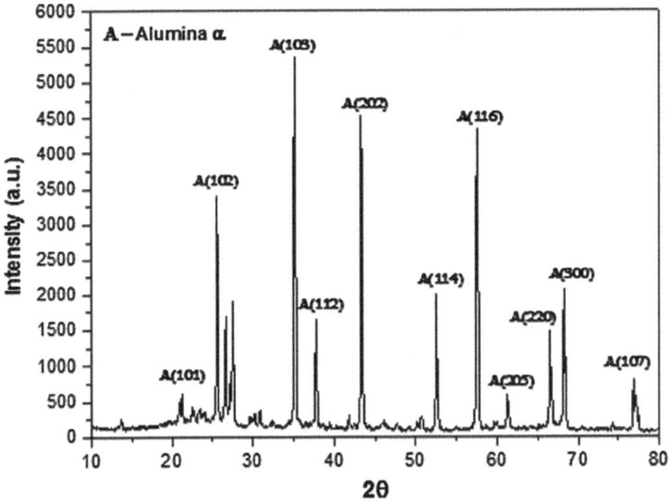

Fig. 4.9 Result of XRD for an α-alumina membrane

References

Arthanareeswaran, G., Devi, T.K.S., Mohan, D.: Development, characterization and separation performance of organic–inorganic membranes. Part II: effect of additives. Sep. Purif. Technol. **67**, 271–281 (2009)

Arthanareeswaran, G., Devi, T.K.S., Raajenthiren, M.: Effect of silica particles on cellulose acetate blend ultrafiltration membranes: part I. Sep. Purif. Technol. **64**, 38–47 (2008)

Arthanareeswaran, G., Thanikaivelan, P.: Fabrication of cellulose acetate–zirconia hybrid membranes for ultrafiltration applications: performance, structure and fouling analysis. Sep. Purif. Technol. **74**, 230–235 (2010)

Biron, D.S., Poletto, P., Duarte, J., Zeni, M., Bergmann, C.P., Santos, V.: Preparation and characterization of PA66/Alumina composite membrane. Mater. Res. **18**(4), 748–755 (2015)

Bowen, D.K., Tanner, E.B.K.: High resolution X-ray diffractometry and topography. Taylor & Francis, London, Bristol (2005)

Brakel, J.V.: Mercury porosimetry: state of the art. Powder Technol. **29**(1), 1–12 (1981)

Bruggen, B.V., Vandecasteele, C.: Modelling of the retention of uncharged molecules with nanofiltration. Water Res. **36**, 1360–1368 (2002)

Brunauer, S., Emmett, P.H., Teller, E.: Adsorption of gases in multimolecular layers. J. Am. Chem. Soc. **60**, 309–319 (1938)

Callister, W.D., Rethwisch, D.G.: Materials science and engineering: an introduction. Wiley, London (2009)

Calvo, J.I., Peinador, R.I., Prádanos, P., Palacio, L., Bottino, A., Capannelli, G., Hernández, A.: Liquid–liquid displacement porometry to estimate the molecular weight cut-off of ultrafiltration membranes. Desalination **268**, 174–181 (2011)

Chen, H., Jia, X., Wei, M., Wang, Y.: Ceramic tubular nanofiltration membranes with tunable performances by atomic layer deposition and calcination. J. Membr. Sci. **528**, 95–102 (2017)

Cullity, B.D., Stock, S.R.: Elements of X-ray diffraction. Pearson Education Limited, London (2014)

Damas, S.B., Miranda, M.I.A., Clar, M.I.I., Roca, J.A.M.: Application of tubular ceramic ultrafiltration membranes for the treatment of integrated textile wastewaters. Chem. Eng. J. **192**, 211–218 (2012)

Echlin, P.: Handbook of sample preparation for scanning electron microscopy and X-ray microanalysis. Springer Science, New York (2009)

Gitis, V., Rothenberg, G.: Ceramic membranes: new opportunities and practical applications. Wiley-VCH Verlag GmbH & Co. KgaA, Weinheim (2008)

Goldstein, J., Newbury, D., Joy, D., Lyman, C., Elchlin, P., Lifshin, E., Sawyer, L., Michael, J.: Scanning electron microscopy and X-ray microanalysis. Springer Science, New York (2003)

Ho, W.S.W., Sirkar, K.: Membrane handbook. Van Nostrand Reinhold, New York (1992)

Li, K.: Ceramic membranes for separation and reaction. Wiley, London (2007)

Mallada, R., Menéndez, M.: Inorganic membranes: synthesis. Characterization and Applications. Elsevier Science, Amsterdam (2008)

Masuda, H., Higashitani, K., Yoshida, H.: Powder technology handbook. Taylor & Francis, London, New York (2006)

Pecharsky, V.K., Zavalij, P.Y.: Fundamentals of powder diffraction and structural characterization of materials. Springer Science, New York (2005)

Porter, M.C.: Handbook of industrial membrane technology. Noyes Publications, New Jersey (1990)

Shang, R., Goulas, A., Tang, C.Y., Serra, X.F., Rietveld, L.C., Heijman, S.G.J.: Atmospheric pressure atomic layer deposition for tight ceramic nanofiltration membranes: Synthesis and application in water purification. J. Membr. Sci. **528**, 163–170 (2017)

Tarleton, S.: Progress in filtration and separation. Elsevier Science, San Diego (2015)

Watt, I.M.: The principles and practice of electron microscopy. Cambridge University Press, Cambridge (1997)

Zeidler, S., Puhlfür, P., Kätzel, U., Voigt, I.: Preparation and characterization of new low MWCO ceramic nanofiltration membranes for organic solventes. J. Membr. Sci. **470**, 421–430 (2014)

Chapter 5
Applications of Ceramic Membranes

This chapter presents some of the possible applications shown by ceramic membranes. The following areas will be given emphasis:

- food Industry;
- water treatment;
- treatment of wastewater;
- biotechnology.

Membrane separation processes feature a wide range of applications. The use of a particular membrane in a certain type of application will depend on its properties, its efficiency, and its interaction with the components of the fluid to be treated.

5.1 Food Industry

The application of membrane separation processes in the food industry has increased dramatically since 1980. This is a great space conquest achieved by the membranes, since many food manufacturers are said to be conservative in approaching new techniques and technologies. In fact, between the 1960s and 1980s, the membrane technology was seen, by the food industry, as a mix of interests and questions.

Regulatory agencies, such as the Food and Drug Administration (FDA) of the United States, and the U.S. Department of agriculture, did not know how to handle the new technology, resulting in a reluctance by the food industry. This explains, for example, that although technology has been used since 1969 by France on cheese industry, only in 1980 the U.S. allowed the insertion of membrane technology as traditional methodology on cheese industry. Currently, few large-scale cheese operations do not require the use of membranes in the process of cheese, whey treatment, recovery of brine, or others.

D. da Silva Biron et al., *Ceramic Membranes Applied in Separation Processes*,
Topics in Mining, Metallurgy and Materials Engineering,
DOI 10.1007/978-3-319-58604-5_5

Nowadays, the acceptance of membrane technology for the food industry is due to several reasons, which are given as follows:

- advances in materials used in membrane production, resulting in more robust membranes and most appropriate sanitation standards and cleaning regimes used by industry;
- improvement of design of membrane modules;
- better understanding of the phenomena of fouling, the main aggravating in the process;
- higher restriction and supervision of the agencies that control the sanitary quality of the food, therefore, greater awareness of the managers of food industry.

5.1.1 Juice Industry

The process of production of fruit juice, including membrane processes, is shown in Fig. 5.1 (Peinemann et al. 2010).

In general, the flow of production in the fruit industry begins with the crunch of the fruits in a certain size and with uniform particles and then a press of the crushed fruit is performed to extract the juice. The process of traditional fining consists on long retention time in tanks, followed by filtration through earth of diatoms, and

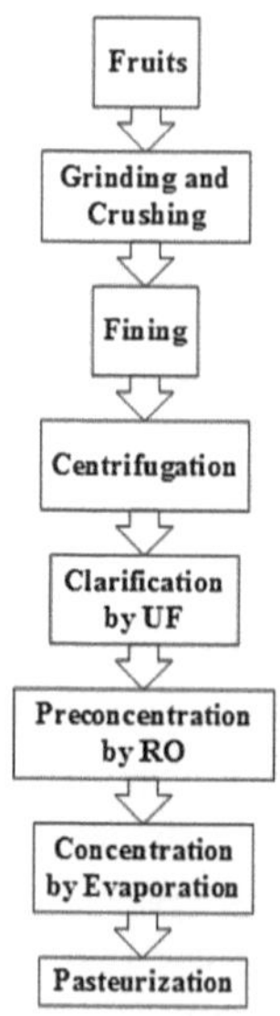

Fig. 5.1 Membrane process in fruit juice production (Peinemann et al. 2010)

requires large amounts of enzymes, gelatin, and other chemicals. After emptying, fruit juice is concentrated to reduce the costs of transport and storage. A common way to concentrate fruit juice is using an evaporator combined with an aroma recovery unit, concentrating the original juice.

Since 1970, the use of UF membranes on clarification of fruit juices such as apple, grape, pineapple, and orange, proved to be a replacement for the traditional process of fining and filtering from both economic and qualitative point of view. The UF process removes suspended solids and other high molecular weight solid, and the filtered juice offers excellent quality with low turbidity. Thus, the UF process replaces the step of fining on a traditional process. In order to achieve high productivity, high capacity, and excellent quality, enzymatic treatment and a proper pre-filtration should be performed before using the UF (Peinemann et al. 2010). In the steps that include processes with membranes, both polymeric and ceramic membranes are used (Mirsaeedghazi et al. 2010; Emani et al. 2013). Ceramic membranes are usually used in tubular shape in tubular modules. Currently, a variety of fruit juices (apple, pear, peach, orange, grape, pineapple, kiwi, strawberry, carrot, etc.) are processed using ceramic membranes (Pabby et al. 2009). However, this type of module is associated with low density of packaging and membrane replacement costs.

5.1.2 Beer Industry

After tea, beer is the most consumed beverage in the world (Peinemann et al. 2010). The beer industry is continually challenged to meet the high standards required for the production of quality beer, with unique flavor and quality while maintaining relatively low production costs and minimizing waste disposal. For the modern consumer, highly demanding, purity, and freshness are extremely important features of a good beer, and that is why the beer industry is permanently looking for new technical solutions to enable a delivery of a product that meets the expectations of the customers (Pabby et al. 2009).

The conventional process of beer, the brewery, starts with the step of malt using hot water to produce must, a sweet, thick liquid. The must is then passed to the wort boiler in which it is cooked for up to 2 h, and later given clarification and cooling. The clarified must is cooled and mixed with yeast and passed to the fermentation tanks where yeast converts sugar from the grain into alcohol and produces the beer by fermentation. Before being transferred to the tanks of bright beer, beer is usually clarified. The finished beer can then be well filtered and pasteurized before bottling. In the case of de-alcoholization, the removal of alcohol occurs before the clarification of beer (Peinemann et al. 2010).

Membrane separation technology is becoming increasingly attractive in the processing steps of the beer industry, with emphasis of applications ranging from beer and whitening cold sterilization until the tank waste retrieval, removal of the alcohol, and water and effluent treatment. The main factors that affect the efficiency of separation and membrane separation performance of beer are the beer composition, characteristics of the membrane, process parameters, and the fouling of membrane. As beer is produced from natural ingredients, its composition presents a wide variety of chemical compounds and, for the most parts has great influence on the membranes and separation process. Carbohydrates, like pentoses and β-glucans, proteins, and protein-polyphenol complex, are particularly important in membrane filtration of beer, as they are responsible for the fouling of the membrane, which has negative consequences on the flow and quality of the filtered beer (Pabby et al. 2009)

In the traditional process of production of beer, the clarification after fermentation and maturation is often achieved by a separator followed by filtration using diatomaceous earths, a process that is associated with the handling and disposal of powder, as well as large quantities of effluents. To overcome these problems, the MF technique crossflow was adopted with plate and frame modules to remove the yeast, microorganisms, and the haze without affecting the taste of beer (Peinemann et al. 2010).

The RO is often used with the purpose of de-alcoholization of beer, as light beer has often been consumed. The dealcoholization is divided into the following stages (Peinemann et al. 2010):

(a) pre-concentration: in this step, beer is separated into two streams. The chain of permeated contains water and alcohol and the retentate contains concentrated beer;
(b) diafiltration: addition of desalinated and deoxygenated water to balance the removal of volume with permeated combined with continuous removal of water and alcohol with the permeated;
(c) adjustment of alcohol: fine-tuning of flavor and alcohol content by addition of desalinated and deoxygenated water;
(d) posttreatment: to balance the loss of flavor due to the removal of the alcohol taste, components such as hops and syrups are added to the dealcoholized beer.

To guarantee the quality of beer, all the steps are performed at temperatures ranging from 7 to 8 °C or lower. After the dealcoholization, the clarification is done and, finally, the bottling (Peinemann et al. 2010).

5.1.3 Wine Industry

The production of wine is still a very traditional process in which filtering is a subject which is often discussed, especially for prestigious wines. However, it can be said that nowadays the fining and filtering have been accepted by winemakers as useful tools to make a better wine (Pabby et al. 2009). The traditional wine-making process begins with the pressing of grapes, followed by the correction of the must, if necessary. The pressing grape juice is centrifuged and transferred to fermentation tanks, where the fermentation process takes place with the presence of yeasts. When the fermentation is complete, a fraction of the wine yeast is removed and the wine is stored in barrels for aging. After aging, the wine is clarified, the tartar is stabilized and filtered and, finally, bottled. Membrane separation processes can replace several of the different stages of separation involved in traditional wine production, as shown in Fig. 5.2 (Peinemann et al. 2010).

In the traditional process of fining, after fermentation, it is often performed with the steps of centrifugation and filtration with diatomaceous earth. Microfiltration membranes can be used to eliminate the number of steps combining clarification, stabilization, and sterile filtration in a continuous operation. Often, MF membranes with pore size of 0.2–0.45 μm are used in clarification of white wine and membranes with pores between 0.45 and 0.65 μm for red wine filtration (Peinemann et al. 2010). However, some studies recommend the use of membranes with pores 0.1 μm for white wines and 0.2–0.4 μm for red wines. Clarification of wine can be performed both by polymeric membranes or ceramic membranes. Most ceramic membranes' providers entered the market of wine filtering with products adapted to the specificities of each type of wine (Pabby et al. 2009).

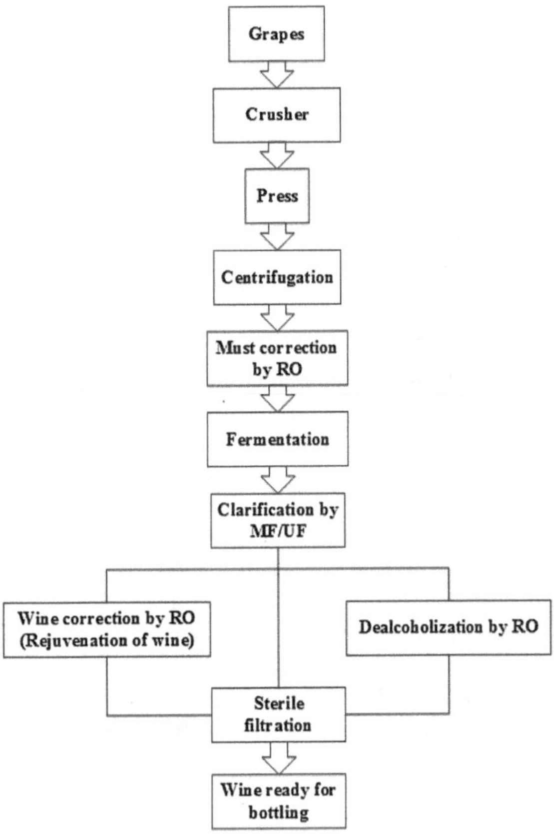

Fig. 5.2 Wine production processes, including those involving membranes (Peinemann et al. 2010)

As well as for other drinks, the main aggravating factor during the filtration of wine is the fouling. This phenomenon is caused by the chemical compounds present in the wine, like polysaccharides and polyphenols.

5.1.4 Dairy Industry

Raw milk at dairy industry is processed with the purpose of obtaining diverse products. The processes include pasteurization, homogenization, separation, and dehydration. The pasteurization process has the advantage of preventing the transmission of diseases, reducing the bacterial load of milk, eliminating undesirable bacteria, prolonging the shelf time of dairy products, and is beneficial to the technology and quality of dairy products. However, this process reduces or eliminates lactic acid bacteria that are beneficial (act as natural yeast), alters the taste of the milk, and causes denaturation of milk. The homogenization process aims to attain a better distribution of milk fat. In separation, the cream and skim milk are divided for the production of butter, cheese, and skim milk. The dehydration process aims the production of hard cheese and milk powders to prolonged storage (Gitis and Rothenberg 2016; Souza et al. 2016).

Since 1960, dairy industry has been using membrane separation processes for clarity, focus, and splitting a variety of dairy products. Membrane processes have enabled the production of refined proteins and their commercial use, which converted the cheese production waste into a valuable product. The integration of membrane separation processes in the dairy industry, based on size of pore and MWCO, allowed the modification of milk by separation and clarification processes in other high quality by-products. The processes of MF, UF, NF, and RO are often used, which allow to separate thoroughly all the main components of the milk. These processes with membranes are used in concentration (NF/RO), fractionation (UF), and filtration (MF) (Scott 1998; Peinemann et al. 2010). Figure 5.3 shows the main membrane separation processes used to obtain the various constituents of the raw milk.

The MF is mainly used in removing bacteria and fats. The MF plants can achieve a reduction of bacteria and spores of more than 99.99% depending on the charge present in the feed flow. The UF is the main process used to increase the protein content of the serum separated and pasteurized. The NF and RO combined can be used for the concentration of soluble proteins and lactose. Soluble proteins and dairy are the most important components of whey. Proteins are separated from milk serum, and can be used in the manufacture of human diet products with specific properties. On the other hand, lactose is used as a supplement in infant formulas and excipients for pharmaceutical products, and can also contribute to the color and flavor of bakery products and pastries (Gitis and Rothenberg 2016).

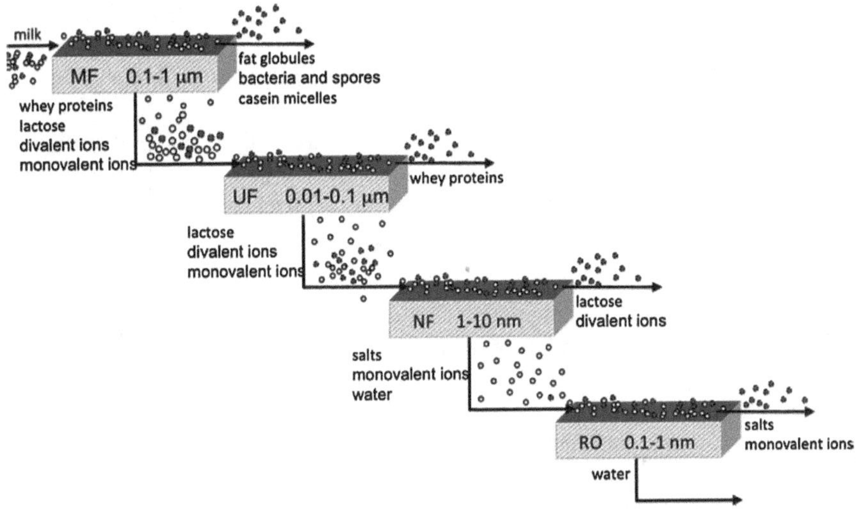

Fig. 5.3 Main processes of separation membranes used in the separation of raw milk (Gitis and Rothenberg 2016)

5.2 Water Treatment

5.2.1 Water Purification

The water is known by all of us as an essential element of life and being essential for survival in simple to complex organisms. The human need for water is not only limited to the consumption, but also the various possible uses of it. Thus, the capture, transport, and storage of the water arose as a result of the increase in consumption resulting from the development of the community, while the treatment, although in its infancy, was born of disgust of man by the aesthetic aspect of the water and developed as a result of the growth of pollution.

Boiling water and water-based drinks were and continue to be the most effective methods in water treatment. The boiling point of water is extremely easy to perform in smaller scale (at home), and is able to eliminate any type of pathogen without the need for sophisticated and expensive equipment. Another ancient and effective method is immersion of different substances such as pearl barley, onions, and wax. Still, since the Roman Empire, the use of silver was carried out in order to eliminate the pathogens present in water, although they could not understand the phenomenon that occurred. During this same period, wealthy families use ceramic filters to purify the water for consumption. Materials such as sand, soil, wool, bread crumbs, or textiles were also used for this purpose (Gitis and Rothenberg 2016).

Natural water purification production for public water supply has as main function the raw water tributary to the station to current applicable standards established by regulatory bodies. In this way, in a water treatment plant for drinking, unit operations and processes are in place that will change the quality characteristics of the raw water, making it exempt (or with low concentrations) of hazardous and noxious substances to human health and may in this way be distributed to the population without risk for the same.

With the aim of limiting the occupied area, decreasing the volume of structures, reducing the extent of channels and pipes, as well as ease of operation and lower costs, the water treatment plants are designed in compact form.

The processes used in water treatment, in general, have the following steps, in the following order:

(a) water abstraction and adduction to the water treatment plant (WTP);
(b) tributary flow measurement by a measurement system, for example, Parshall flume;
(c) mechanical or hydraulic coagulation to promote contact of coagulant (e.g., aluminum sulfate and ferric chloride) with colloids in water;
(d) mechanical or hydraulic flocculation to promote the formation and increase of flakes. In some cases, polyelectrolyte is used as auxiliaries in coagulation;
(e) conventional or high sedimentation rate with low gradient mixing values to promote the removal of flakes formed in the previous step;
(f) conventional filtration (slow or fast filters) for the removal of particles not established in the previous step or are of low mass;
(g) Chemical (e.g., use of chlorine) or physical (e.g., use of ultraviolet radiation) disinfection process to the elimination of microorganisms. Some countries, especially those in development, which have water storage systems in homes, use the disinfection with a chemical agent that provides residual, promoting longer disinfection;
(h) and, finally, storage and distribution to the population.

Similar to polymeric membranes, ceramic membranes are used in water purification processes. However, because of the disadvantages of ceramic membranes, such as the high cost and low efficiency for certain contaminants, ceramic membranes were put aside, giving space for the growth of polymeric membranes on the market (Gitis and Rothenberg 2016).

An alternative to conventional filtration is the use of MF membranes. This process, in general, enables to remove particles such as sand, clay, giardia, cryptosporidium, algae, and some bacteria, but does not remove viruses. On the other hand, the UF can be used in the removal of bacteria and most viruses, however, it recommends the use of chemical disinfection, such as chlorine. The UF can be operated completely automated and does not require the use of chemicals and produces quality water (Yang and Kim 2009).

NF membranes possess greater sensitivity to MF and UF membranes. These can eliminate any type of bacteria or viruses, in addition to carbonates, bicarbonates, bivalent ions, and compounds responsible for the toxicity of water (for example, hexavalent chromium) (Dasgupta et al. 2015 and Chon and Cho 2016).

The RO process generally only allows the passage of water, retaining all the compounds present in the water. The feed is subjected to high pressures which force the flow of water through the pores smaller than 1 nm, retaining all salts present in the water and reaching conductivity below 1 μS/cm. In cases involving RO membranes, the use of pretreatment is essential. Many times, the treatment of raw water is performed by MF/UF membranes followed by RO membranes. The preliminary treatment of the water, before RO, aims to preserve the quality of the membrane, as this has a lot of sensitivity, but may decompose along the process (Gu et al. 2016).

5.2.2 *Water Desalination*

On Earth, 97.5% of the water is considered salt water and 2.5% fresh water, but only 0.4% is available for consumption. Due to this and the difficulty of access, many countries resort to the treatment of salt water. Different technologies are used for the desalination of water potability of the water. These processes include thermal desalination plants, use of solar energy, desalination by RO, and others. The conventional thermal desalination leads to emission of harmful gases and toxic to the environment, apart from consuming a lot of energy. The use of solar energy has the advantage of using a renewable source and is ecologically friendly, but still not usable due to the limitation in efficiency and the high costs of equipment used for the desalination system. Most of the times, the water supply to the growing population is through the use of desalination systems for RO (Chandrashekara and Yadav 2017). The increasing use of RO on desalination of water is mostly due to the high production of water with a low additional value. A filtration system with RO membranes use basically four main components: pretreatment, high pressure pumping system, RO membrane modules, and posttreatment, the latter if necessary (Ali et al. 2017).

The commercial interest in RO technology has globally increased, over the years, due to the continuous process improvements, which lead to significant reductions in costs. Still, these advances include the development of materials and design of membrane modules, drawings of processes, pretreatment, and energy recovery. However, the greatest efficiency gains have emerged from the improvement of materials used in obtaining RO membranes. The structure, the material, and the morphology of the membranes were modified to improve functions (permeability and selectivity) and applicability (mechanical, chemical, and biological stability) (Lee et al. 2011).

The use of ceramic membranes in the production of domestic water is rare, but its robustness in cases has attracted the attention of researchers for applications in MD and pervaporation. Given the potential water desalination in the oil field, and being motivated by molecular dynamics simulation results, showing 100% ion rejection by the perfect all-ZK-4 zeolite membranes, investigations have been carried out to understand the mechanisms of separation and the feasibility of application of ceramic membranes (Lee et al. 2011).

5.3 Wastewater Treatment

Residual water can be understood as a liquid that contains residues in the form of organic matter, suspended solids, and other contaminants, originated from human activities. This liquid when not properly treated ends in polluting the water sources, causing impairment of the quality of aquatic life, and harming them. In general, residual water can be divided into domestic or industrial.

From the point of view of human health and environmental impacts, residual water treatment technology is extremely advanced. In general, considering the volumes, the wide range of existing contaminants, and rigorous objectives involved in residual water treatment processes, the capacity of society to manage its liquid waste so effectively is remarkable. However, over the past 20 years the improvement of the techniques of recovery of resources (water, nutrients and energy) available in domestic waste water was necessary (Metcalf and Eddy 2014).

Domestic residual water is the most abundant type of residual water that falls into the category of waste streams of low resistance, characterized by low resistance and high organic content of organic matter. The treatment of domestic waste water has the potential to become producers of renewable energy, by converting the chemically bound energy content of pollutants in urban residual water to a useful energy carrier. Therefore, there is a need for choice of a suitable energy recovery technology that can convert the existing energy in residual water to a renewable energy source (Ozgun et al. 2013; Metcalf and Eddy 2014).

Another relevant aspect is the scarcity of fresh water, which has been worsened due to increased population density and contamination of water resources. A way to increase water resources in regions with water scarcity is reusing treated residual water in the agricultural, industrial, and urban use. Currently, the liquid waste from urban areas is dumped on the environment and generally is considered as waste. To work around this, various treatments can be applied for better water quality (Sert et al. 2017).

Currently, mostly in developed countries, technologies prevail as the conventional activated sludge systems, coagulation/flocculation followed by sedimentation, and others. However, the membranes over the years, with the gain of reliability, are being gradually incorporated into residual water treatment processes. An example is the use of systems of membrane bioreactor (MBR), which is currently found in several residual water treatment plants. A reuse of domestic residual water may be done by MBR process, combining biological treatment and membrane filtration. Commonly, UF or MF membranes are used in MBR systems to separate aqueous phase treated solids and thus provide a solid and free permeated high quality (Sert et al. 2017). Both ceramic membranes and polymeric membranes are used for this purpose. Niwa et al. evaluated a MBR system with ceramic membranes in residual water treatment. In this work, in full scale, a removal of chemical oxygen demand (COD) of 91% was achieved. A-alumina membranes used featured pore size of 0.1 μm. Still, the MBR plant with the upflow anaerobic sludge blanket (UASB) presented a 0.76 kWh energy consumption mm^{-3} (Niwa et al. 2016).

Although MBR processes seem to be an alternative solution, providing a high-quality effluent for discharge, it is still not possible to use this water directly for irrigation or in the process. The MBR effluent may still contain dissolved salts that must be removed before the water can be reused. Therefore, it would require an additional treatment (Sert et al. 2017).

Various activities, including the tannery industry, abattoirs, electroplating industry, wine industry, and others, present a high load of pollutants in residual water, which cause a high depreciation of the springs when improperly discarded. The conventional effluent treatment processes generated often do not reach the requirements of current legislation, making necessary the use of other technologies for additional treatment. The environmental appeal, currently, fortifies the search for technologies that aim not only treatment but also the insertion of the treated effluent to process, diminishing or avoiding total discard. At this point, the membrane technology aims to reach these goals.

The tanning industry is known for high load of organic materials and by the presence of heavy metals. Many industries, this follow-up, present difficulties in treatment, mainly due to the presence of nutrients and various metals. Bhattacharya et al. used a membrane separation system, combining the process of MF (ceramic membrane) to the RO process (polymer membrane). In this study, the tannery effluent, originally retrieved from a tannery industry located in Kolkata, presenting high load of organic matter, nutrients, solids, and metals. After the treatment of residual water by combined MF and RO systems, it was possible to check almost 100% efficiency of COD removal, 99.95% of phosphorus removal, and 99.98% chromium removal. These results demonstrate the effectiveness of membrane separation processes when applied in the treatment of residual water, even when the effluent has high load of contaminants (Bhattacharya et al. 2013).

5.4 Biotechnology

In general, biotechnology is a set of knowledge that allows the use of biological agents (such as living organisms and enzymes) to obtain a product with additional value. Biotechnology involves several sectors such as energy generation, industrial products (such as ethanol), environment (bioremediation), agriculture (for example, biofertilizers and genetic selection of plants), and others. Despite the wide definition and diverse applications in this area, for this book, biotechnology will be limited in areas that involve the membrane separation processes only.

In biotechnology, the membranes are often used in or fermenters and bioreactors to obtain, for example, alcohol or the concentration of enzymes. Membrane separation processes are increasingly being inserted in fermenters or bioreactors. Primarily, they are used for carrying gases, such as oxygen and carbon dioxide, and to selectively transfer the culture out of the reaction tank (Turner 1991; Crespo and Böddeker 1993).

The use of membranes associated with small-scale systems is included within a small but lucrative industry. These bioreactors have been used mainly for cultivating cells of mammals, plants, and insects on a small scale by pharmaceutical and biotechnological companies. Systems with associated membranes bioreactors, mostly, are still in laboratory scale. The goal is to quickly get enough amount of the desired product to test its effectiveness or other features. In biorefineries, bioethanol production from lignocellulosic agricultural residue presents an alternative method to burning fossil fuels. However, due to some technological barriers, bioethanol production from lignocellulosic waste has not been produced on a large scale yet. To achieve the goal, bioethanol production technology, which is economically viable and includes pretreatment, enzymatic hydrolysis, fermentation, and dehydration, needs to be developed. In the bioethanol industry, the membrane is generally used in purification, concentrating the ethanol produced. But it can be used in recycling of cellulase enzymes, cell recycling, and removal of inhibitory products (Lipnizki 2010 and Saha et al. 2017).

Fermentation of lignocellulosic biomass for ethanol fuel faces two major difficulties: the cellulose crystalline structure that resists cell enzymatic hydrolysis, cellulose–lignin pulp association hampers the enzymatic attack to the cellulose. Still, the cost of the enzyme is 20% of the total cost of production of ethanol and 50% of the total cost of the complete hydrolysis process. One way to reduce the cost of ethanol production from lignocellulosic biomass is to recover, recycle, and reuse the enzyme as well as the specific activity of the enzyme. To this end, the recovery and recycling of the enzyme require separation processes. In this case, the membrane separation processes are the most suitable processes. Membranes can retrieve the enzyme, retain their catalytic activity, recycle, and reuse the enzyme in a new procedure of hydrolysis (Saha et al. 2017).

Figure 5.4 shows, in general, the process of obtaining ethanol that includes the potential use of membranes.

Bioethanol production can also be performed based on starch. The typical raw materials of starch come from maize, wheat, cassava, and potatoes. In addition, two different concepts to produce starch-based bioethanol are currently carried out on an industrial scale: separate hydrolysis and fermentation (SHF) and simultaneous hydrolysis and fermentation (SFF) (Lipnizki 2010).

The MF and UF are then used in recycling and cell concentration. On the other hand, the RO is used as a condensate polisher system. Still, the pervaporation, as promising technology for the removal of ethanol in situ fermentation systems, contributes to the reduction of ethanol inhibition. Compared to traditional techniques, such as gas extraction and adsorption, the pervaporation is simple, highly selective, and not toxic to microorganisms' fermentation. On pervaporation, due to the nature of the ethanol–water vapor liquid equilibrium, the ethanol prefers to permeate through the pervaporation membranes at low concentrations of ethanol (He et al. 2012).

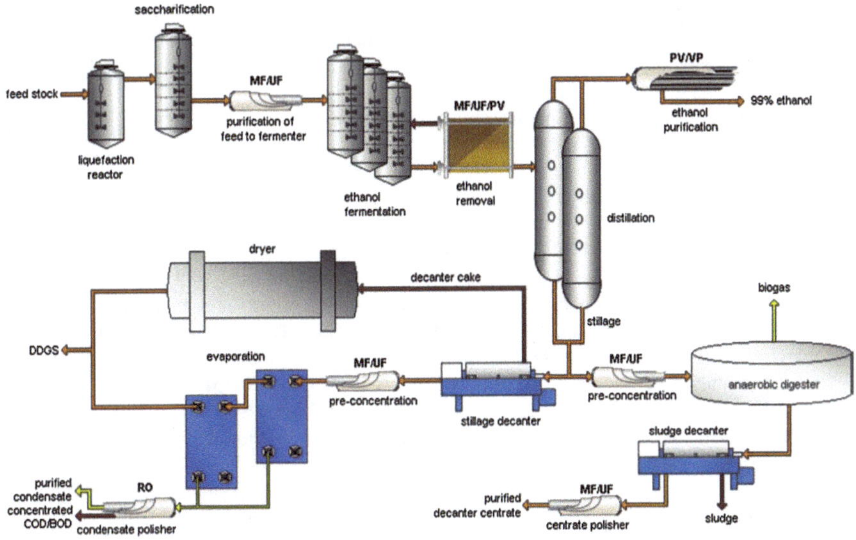

Fig. 5.4 Ethanol production process (SHF) with potential application of membranes (Lipnizki 2010)

References

Ali, E.S., Alsaman, A.S., Harby, K., Askalany, A.A., Diab, M.R., Yakoot, S.M.E.: Recycling brine water of reverse osmosis desalination employing adsorption desalination: a theoretical simulation. Desalination **408**, 13–24 (2017)

Bhattacharya, P., Roy, A., Sarkar, S., Ghosh, S., Majumdar, S., Chakraborty, S., Mandal, S., Mukhopadhyay, A., Bandyopadhyay, S.: Combination technology of ceramic microfiltration and reverse osmosis for tannery wastewater recovery. Water Resour. Ind. **3**, 48–62 (2013)

Chandrashekara, M., Yadav, A.: Water desalination system using solar heat: a review. Renew. Sustain. Energy Rev. **67**, 1308–1330 (2017)

Chon, K., Cho, J.: Fouling behavior of dissolved organic matter in nanofiltration membranes from a pilot-scale drinking water treatment plant: an autopsy study. Chem. Eng. Journ. **295**, 268–277 (2016)

Crespo, J.G., Böddeker, K.W.: Membrane Processes in Separation and Purification. Springer, Curia (1993)

Dasgupta, J., Mondal, D., Chakraborty, S., Sikder, J., Curcio, S., Arafat, H.A.: Nanofiltration based water reclamation from tannery effluent following coagulation pretreatment. Ecotoxicol. Environ. Saf. **121**, 22–30 (2015)

Emani, S., Uppaluri, R., Purkait, M.K.: Preparation and characterization of low cost ceramic membranes for mosambi juice clarification. Desalination **317**, 32–40 (2013)

Gitis, V., Rothenberg, G.: Ceramic Membranes: New Opportunities and Practical Applications. Wiley-VCH Verlag GmbH & Co. KgaA, Weinheim (2016)

Gu, H., Rahardianto, A., Gao, L.X., Christofides, P.D., Cohen, Y.: Ultrafiltration with self-generated RO concentrate pulse backwash in a novel integrated seawater desalination UF-RO system. J. Membr. Sci. **520**, 111–119 (2016)

He, Y., Bagley, D.M., Leung, K.T., Liss, S.N., Liao, B.-Q.: Recent advances in membrane technologies for biorefining and bioenergy production. Biotechnol. Adv. **30**(4), 817–858 (2012)

Lee, K.P., Arnot, T.C., Mattia, D.: A review of reverse osmosis membrane materials for desalination—development to date and future potential. J. Membr. Sci. **370**(1–2), 1–22 (2011)

Lipnizki, F.: Membrane process opportunities and challenges in the bioethanol industry. Desalination **250**(3), 1067–1069 (2010)

Metcalf and Eddy: Wastewater Engineering: Treatment and Resource Recovery. McGraw-Hill Education, Boston (2014)

Mirsaeedghazi, H., Emam-Djomeh, Z., Mousavi, S.M., Aroujalian, A., Navidbakhsh, M.: Clarification of pomegranate juice by microfiltration with PVDF membranes. Desalination **264** (3), 243–248 (2010)

Niwa, T., Hatamoto, M., Yamashita, T., Noguchi, H., Takase, O., Kekre, K.A., Ang, W.S., Tao, G., Seah, H., Yamaguchi, T.: Demonstration of a full-scale plant using an UASB followed by a ceramic MBR for the reclamation of industrial wastewater. Biores. Technol. **218**, 1–8 (2016)

Ozgun, H., Dereli, R.K., Ersahin, M.E., Kinaci, C., Spanjers, H., van Lier, J.B.: A review of anaerobic membrane bioreactors for municipal wastewater treatment: integration options, limitations and expectations. Sep. Purif. Technol. **118**, 89–104 (2013)

Pabby, A.K., Rizvi, S.S.H., Sastre, A.M.: Handbook of Membrane Separations: Chemical, Pharmaceutical, Food, and Biotechnological Applications. Taylor & Francis Group, Boca Raton (2009)

Peinemann, K.V., Nunes, S.P., Giorno, L.: Membrane Technology, Volume 3: Membranes for Food Applications. Wiley-VCH Verlag GmbH & Co. KgaA, Weinheim (2010)

Saha, K., Maheswari, U.R., Sikder, J., Chakraborty, S., Silva, S.S., Santos, J.C.: Membranes as a tool to support biorefineries: applications in enzymatic hydrolysis, fermentation and dehydration for bioethanol production. Renew. Sustain. Energy Rev. **74**, 873–890 (2017)

Scott, K.: Handbook of Industrial Membranes. Elsevier Science, Oxford (1998)

Sert, G., Bunani, S., Yörükoğlu, E., Kabay, N., Egemen, Ö., Arda, M., Yüksel, M.: Performances of some NF and RO membranes for desalination of MBR treated wastewater. J. Water Process Eng. **16**, 193–198 (2017)

Souza, R.R., Bergamasco, R., Costa, S.C., Feng, X., Faria, S.H.B., Gimenes, M.L.: Recovery and purification of lactose from whey. Chem. Eng. Process. **49**(11), 1137–1143 (2010)

Turner, M.K.: Effective Industrial Membrane Processes: Benefits and Opportunities. Elsevier Science, Cranfield (1991)

Yang, H.-J., Kim, H.-S.: Effect of coagulation on MF/UF for removal of particles as a pretreatment in seawater desalination. Desalination **247**(1–3), 45–52 (2009)

Chapter 6
Ceramic Membrane Modules

Membrane module can be understood as a physical unit, which gathers all the elements that participate in the separation process. Membrane module is, basically, composed by membranes, channels where feed goes through, the permeate and concentrate, and by pressure or current support structures applied to the system. The membranes are found inside these modules, where they are protected from impacts or contact with the external environment, and supported by the modules. These modules can be made of various materials, such as polypropylene (PP), polyethylene (PE), polyvinyl chloride (PVC), and stainless steel.

Generally speaking, the modules should be projected as to favor simplicity of handling, ease for cleaning, low dead volume, limit the accumulation of material retained on the membrane, maximize the membrane's area by module volume, and avoid the contamination of the permeated with the concentrated material.

There are six distinctive modules, in which membranes are displayed:

- plate and frame modules;
- spiral wound modules;
- tubular modules, with one or more tube beam;
- perforated block modules;
- hollow fiber modules; and
- rotating disk modules.

The following sub-chapters have the purpose of presenting the main characteristics and differences between the membrane modules. The modules that will be discussed were mentioned previously, but in addition to these, there is a wide range of other modules with different configurations and characteristics, but the most important are among them.

D. da Silva Biron et al., *Ceramic Membranes Applied in Separation Processes*,
Topics in Mining, Metallurgy and Materials Engineering,
DOI 10.1007/978-3-319-58604-5_6

6.1 Plate and Frame Modules

Plate and frame ceramic membranes can be under shape of plate leaves or disks. Both shapes can be packed into module-type plate and frame (Basile et al. 2015). Plate and frame modules are considered the first configurations of modules ever made and commercialized in history. The commercial development of those modules for plate membranes started in the 1960s and is considered the oldest configuration still in use. Its configuration is very similar to a filter press. Generally, plate and frame module consists in multiple units of plate leaves or disks packed together, like a multiple-layers sandwich. According to Fig. 6.1, each unit includes four porous supporting plates, two plate leaves or plate membrane disks and channels that allow feed and permeated passage. Also, the presented example in this figure is one among many others (Ho and Sirkar 1992).

Membranes are sealed to supporting plates by gaskets with locking devices, glue, or are directly attached. The supporting plate is, internally, porous and proportionate, a flow channel through permeated, which is collected next to the supporting plate. Feed is conducted through feed channels, which have contact with the membrane's surface. What is rejected is conducted through a main channel and will go to a feed tank, and go back to the entrance with the feed. The feed channel can have a free path with heights of 0.3–0.75 mm. Higher channel heights are necessary to fluids of high viscosity. It is possible to reach a reduce in energy consume when utilizing larger widths of the feed channel. Channel heights will depend on quantity and size of solids suspended on feed (Ho and Sirkar 1992).

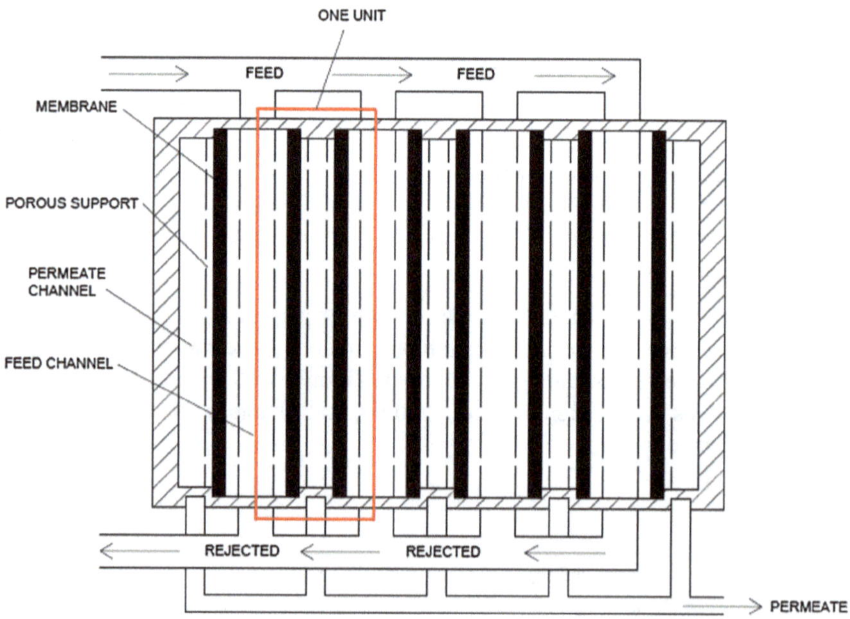

Fig. 6.1 Example of plate and frame module

Modules with plate and frame configuration present advantages such as easy separation of water solids and easy cleaning of membranes' surface. Some specific modules, the ones with crossflow configuration, allow such module to be rotated, allowing shear forces action and reducing fouling. Depending on the application, plate and frame membrane modules can be of easy handling. On the other hand, this system presents some disadvantages. The biggest problem present on this module is its low packing density (100–400 m^2/m^3) (Mulder 1996). It has low efficiency in comparison to other configurations and presents pressure fall. Also, this module configuration presents, in most of the times, high cost. Besides, in this module, the number of Reynolds obtained characterizes flow in the laminar region (Purchas and Sutherland 2002).

The biggest application of such modules in on MBR (Membrane Bio Reactor) processes, due to its capacity of operating easily with high concentration of solids on feed. Another application of these modules is on electrodialysis processes, due to its characteristic of bigger distribution in electric current. Besides, such module can be applied in many other processes, such as in MF, UF, NF, and OI (Hoek and Tarabara 2013).

6.2 Spiral Wound Modules

Spiral wound modules had their incorporation on separation processes with membranes of 1965, as an alternative to plate and frame and tubular modules, popularly used back then. The spiral wound module can be understood as a plate membrane sheet interleaved between feed channels and permeated channels, stating that between two membranes there will be a spacer called permeate carrier. This set is rolled around a hollowed pipe called permeate main channel. Figure 6.2 presents a simplified way a spiral wound module. The elements involved in these spiral modules used crossflow technology, and, due to construction, can be easily created in different configurations with length, diameters, and various membranes' materials. The feed spacer's objective is to proportionate space so that the fluid transits along the membrane's surface, and allow a uniform flow among membranes. Permeate spacers are linked to membranes and connected to the main permeate tube. This spacer's objective is to transfer permeated flux to the main channel (Hoek and Tarabara 2013; Karabelas et al. 2015).

The feed flows through the channels, tangentially, along the length of the permeate main tube. The filtered goes to permeate carrier and then to the permeate main tube. The concentrate, rejected, or retentate will go to the other border (Purchas and Sutherland 2002).

Just as tubular module, the spiral wound presents as characteristic of an operation in turbulent regime. The main application of these modules is in OI processes, and can be applied to MF, UF, and NF processes. The preference of application of these modules on desalination of seawater is due to balance between ease of handling, control of fouling, permeation rate, packing density, such as initial,

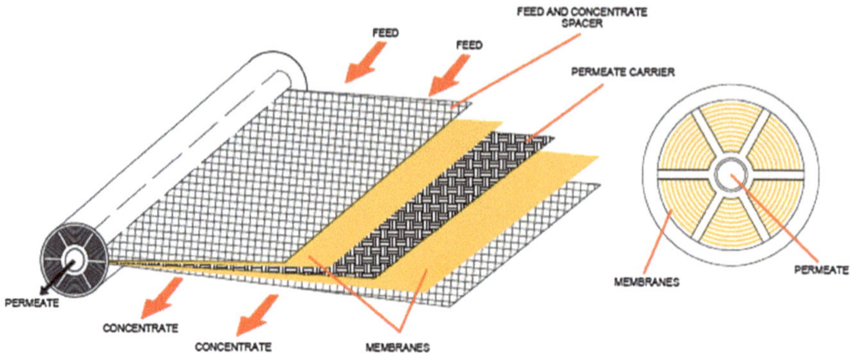

Fig. 6.2 Schematic structure of a spiral wound module

operating, and replacement costs (Wang et al. 2011). According to Porter (1990), reverse osmosis membranes that are used in spiral modules configuration present, in industrial-scale, dimensions around 5.08 and 30.48 cm of diameter and 30.48–152.4 cm of length, with packing density up to 984 m^2/m^3.

Spiral modules present many advantages when compared to plate and frame and tubular modules. The most significate advantage, presented by spiral modules, is the packing density (300–1000 m^2/m^3), which is more than the double when related to plate and frame and tubular modules. This characteristic results in lower area requirements, and, consequently, on a reduction of cost when installing a filtration plant. Another advantage is the possibility of using many configurations with different spacers, membrane types, lengths, and diameters, which allow its usage in various applications. However, these modules present disadvantages such as cleaning difficulty, lower packing density compared to hollow fiber modules, and higher fouling related to tubular modules. The elevated fouling in such membranes is the consequence of difficulty of mechanical cleaning, meaning that such modules cannot be disassembled to the mechanical cleaning. The consequence of this is the increase of the operating pressure and of the frequency of chemical cleaning to remove fouling and scaling formed and, the decrease in the useful life due to the use of chemical agents in the cleaning. Still, this module cannot be used for a fluid with high concentration of suspended solids or with suspended solids of high size (Mulder 1996; Hoek and Tarabara 2013).

6.3 Tubular Modules

Studies with membranes in tube shapes started in 1920; however, its application in industrial scale started in the 1960s. Even though Loeb and Sourirajan, initially, obtained plate membranes of cellulose acetate, they shaped tube membranes, giving origin to a new way of process (Hoek and Tarabara 2013).

The main characteristics of tubular module are as follows:

(a) easy replacement of tubular membranes in an individual way;
(b) easy cleaning of the contaminated surface, as for physical and chemical methods;
(c) high commercialization cost;
(d) high energy consumption per unit of treated liquid;
(e) high feed flow rate helps to reduce the contamination of a membrane;
(f) module's low packing density (20–30 m^2/m^3);
(g) simple pretreatment of feed liquid; and
(h) operates in turbulent regime.

The process which utilizes this configuration allows the presence of big-sized particles on the fluid, as the size of the particle must correspond to a tenth of the channel's height. For example, tubular units with internal diameter of 12 mm allow the presence of particles of up to 1200 μm on the feed fluid.

Figure 6.3 shows a typical tubular module scheme. In general, the tubular module can be understood as one or more tubular membranes disposed in a hollow tubular housing, which has entranced to the feed and exits to the concentrate and permeate. These modules are similar to shell and tube heat exchangers (Scott 1998). Ceramic membranes are the most used in this configuration, due to its stiffness and shape which adequates easily to the system. Ceramic tubular membranes are hollow and straight structures, which have wall thickness of 2–5 mm, total diameter of 10–52 mm, inside diameter of 2–16 mm, and length of 0.2–1.2 m. The tubular membranes can be produced on either mono-channels or multichannels. Multichannel will be discussed in the upcoming chapter. Feed flows internally along the tube, and permeate goes through the selective, intermediate, and supporting layers, getting to the module hull, and passing through the holes disposed in the module.

Tubular modules work through tangential flux and are generally used to process feed flows with high load of dissolved, suspended and with considerable quantity of oil and fat solids. Tubular modules can be used on a high flow speed to minimize the concentration of substances on the membrane's surface. The flow's speed influences directly on energy consumption, as under turbulent conditions the feed

Fig. 6.3 Multichannel tubular module

rating will increase with speed increase. Permeate flow is usually a function of the feed flow velocity and optimization needs to be done between energy consumption and the membrane's superficial area (Pabby et al. 2009).

Modules can be arranged vertically and horizontally. Some systems are operated with additional aeration to the fouling control. Tubular modules give direct hydrodynamic control on the membrane's surface. These modules can be applicated in MF, UF, and NF processes. Tubular modules are widely used on clarification of wines and juices, water treatment and effluents, in the treatment of oily waters, distillation membrane (MD), on gas separation among others (Li et al. 2008; Pabby et al. 2009).

6.4 Perforated Block Modules

This subchapter is about modules that use multichannel ceramic membranes (Fig. 6.4). These modules do not present great differences when compared to modules that use mono-channel membranes. The only difference is the structure in shape of Perforated Block.

The tubular shapes are the most likely to be adopted to inorganic environment, especially to ceramic materials, where the idea is extended to perforated block structures. The process of obtaining perforated blocks consists in perforate a porous ceramic monoclinic block with a determined number of channels, parallel to its length. The monoclinic block is then coated with an active layer of ceramic membrane over the internal surface. It is possible to obtain a wide range of channels and differentiated forms with this type of structure. The shapes range to a total of

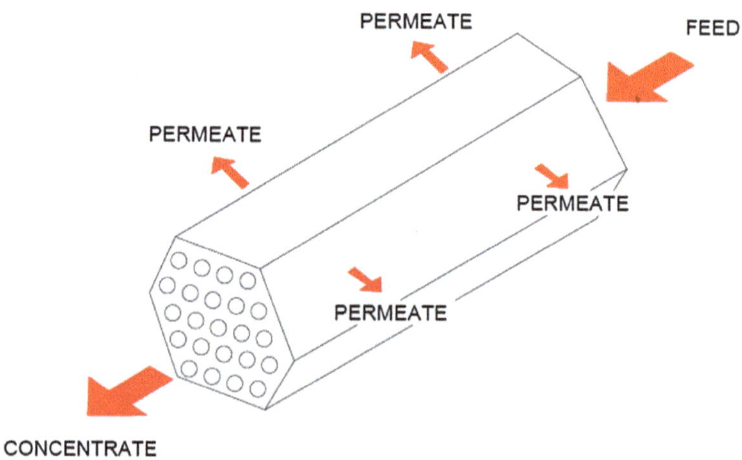

Fig. 6.4 Multichannel tubular membrane

14, with up to 39 individual channels, which offer diameter between 2 and 14 mm (Purchas and Sutherland 2002). The main geometry of perforated blocks modules is circular or hexagonal.

The main advantage of perforated blocks is the increase of the membrane filter area, which allows higher permeate flow. As, for example, if a membrane presents a mono-channel structure with internal diameter of 16 mm and length of 1 m, it will present filter area of 0.13 m^2. This difference corresponds to a value of 61% on the increase of the filter area to a multichannel membrane. As consequence, these perforated modules present an increase on packing density. This difference is due to lower wall thickness presented by these membranes, but it is only possible over higher mechanical resistance that perforated block presents as it is a unique piece. Besides these advantages, the module presents, just as tubular module, ease of cleaning, high feed flow in turbulent regime, among others. However, it has more cost when compared to other modules and needs more energy to work.

6.5 Hollow Fiber Modules

As with many other modules, hollow fiber modules were first developed in the 1960s. Even though this module's configuration is applied mainly to polymeric membranes, as polysulfone and polyacrylonitrile, some papers show researcher's tries on shaping ceramic membranes on hollow fiber for their use in such modules (Hubadillah et al. 2016; Arndt et al. 2016; Wit et al. 2017). According to Hubadillah and collaborators, hollow fiber ceramic membranes present superior characteristics compared to other types of membranes, such as excellent thermal and chemical stability. However, its fragility and high obtaining cost has difficulted its commercial use. The obtainment of these membranes uses a simultaneous process of the method of spinning and phase inversion. After the spinning process, fibers are washed and sintered to high temperatures. After sintering, the supporting layer is covered by a functional ceramic coating layer and sintered another time. This process of obtaining hollow fiber ceramic membranes can be applicated in an infinity of ceramic materials, as alumina, silicon carbide, silicon oxide, among others.

Hollow fiber module can contain from 50 to 10,000 individual fibers (depending on the fiber's and module's diameter), which are attached to both borders by an epoxy resin (Scott 1998). The feed fluid can enter tangentially or perpendicularly to the fibers. The first configuration can be verified in Fig. 6.5. We can see in the figure that the feed goes through all along the fiber filament, and that the concentrate exits by the other border, and the permeate is collected on the module hull. On the second configuration, the entrances are inverted; it means, the feed will enter through the permeate channels and the feed and concentrate exits will be to collect

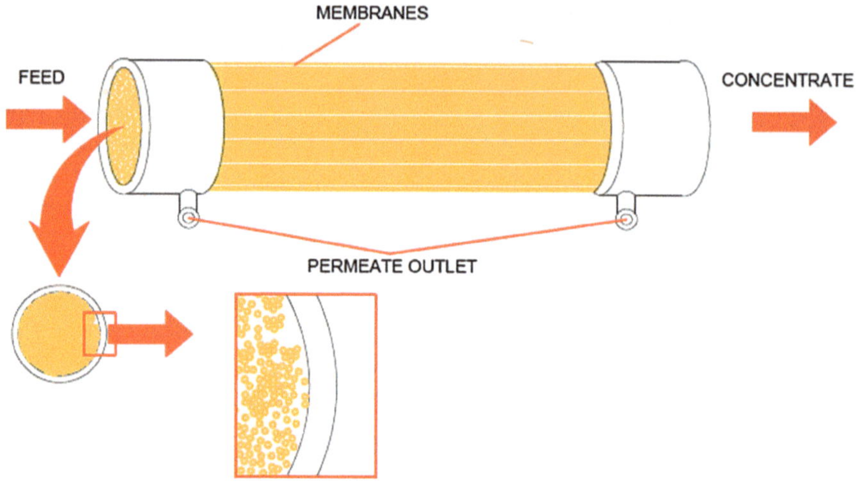

Fig. 6.5 Hollow fiber module

the permeate (Escobar and Bruggen 2011). The main difference in both configurations is the decrease of the polarization layer that occurs when putting the fibers transversely to the feed flow (Pabby et al. 2009). The size of fibers makes backwash indispensable for longer service life of this kind of membrane; however, this module is flexible to the realization of this process.

Hollow fiber membranes are considered ideal on the process of gases separation, due to its high packing density, from 600 to 1200 m^3/m^2, reaching up to 30,000 m^3/m^2 (Mulder 1996; Pabby et al. 2009). Besides the high packing density, hollow fiber modules present great flexibility of fibers, which allows certain configurations that cannot be reached with other membranes. Still, the cost is relatively low to great facilities, the energy cost is low, and the filter systems occupy a relatively small area. However, these modules do not present low cost for small facilities, the membranes are easily damaged, not being able to operate with high rates of suspended solids, and the module assembly is complicated. Besides the presented disadvantages, these modules present high cost of reposition, because the rupture of a single fiber implies the complete reset of the module, and the rupture of a fiber cannot be detected, which implicates in a loss of process quality (Li et al. 2008).

Hollow fiber membranes contain around 1 mm of diameter, while the thickness is from 250 to 500 μm. The feed flow operates at a Reynolds of 500–3000. The hollow fiber modules can be applied in MF, UF, NF, and OI processes. Many authors presented the study of application of these modules on treatment of water and domestic sewage, tertiary treatment of water for reuse, and in the treatment of industrial effluents (Shimizu et al. 1996; Busch et al. 2007; Yuliwati et al. 2011; Govardhan et al. 2017).

6.6 Rotating Disk Modules

The permeate flow decrease, caused by polarization layer and fouling, is the most important and worrying phenomenon during the process of separation by membranes. Many studies worry about the reduce of such phenomenon, making changes in the characteristics of the process and the membrane, as the modification of the surface and structure of membrane, as well as in the module's modification. It is possible to say that the emergence of rotating disk modules can be attributed as an alternative to reduce the phenomenon. When the module hits rotation speed of 15–20 m/s, it generates high shear rates, which provokes the removal of solute and organic matter adhered to membranes (Brou et al. 2002; Bouzerar et al. 2003; Bhattacharjee and Bhattachary 2006).

In this module, Fig. 6.6, the membrane disk is displayed in a rotary hollow shaft, which allows permeate passage. The entrance of feed occurs along points localized between the membranes. The feed fluid, after passing through the membrane pores, is directed through a porous environment (permeate channel), to the main permeated channel. In this system, there is no exit for the concentrate and the cylinder of the system remains under pressure.

When these modules are used in microfiltration of cell or particle suspensions, they produce higher permeate flows than the conventional crossflow filtration, with an excellent transmission of micro-solute due to the lack of particle layer on the membrane. However, these systems are more complex and present a higher cost for construction than more conventional units, and can also consume more energy to rotate disks. Besides, these systems present difficulty on its scheduling and low packing density (Brou et al. 2002).

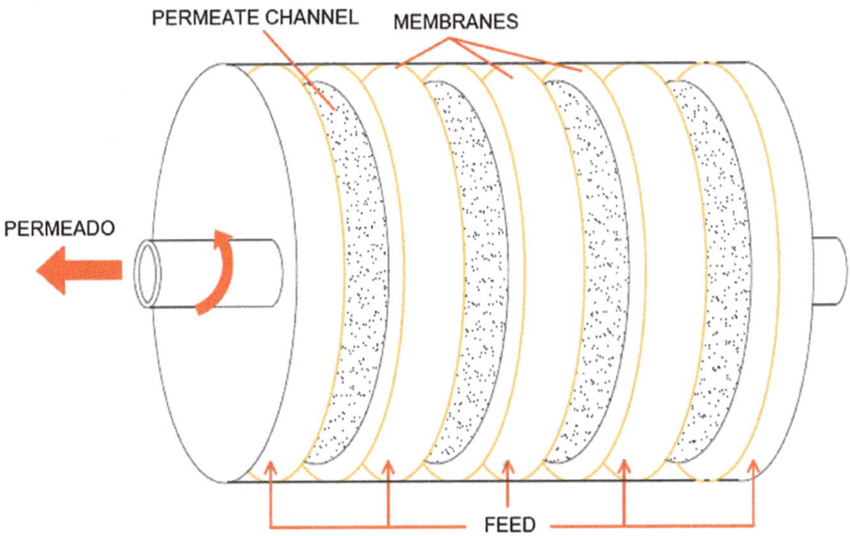

Fig. 6.6 Rotating disk module

In this module, the mass transference occurs by the induction of instability or turbulent flows, being its rate determined by the rotation rate of the internal cylinder, which may vary from 300 to 3000 rpm. In tangential flow systems, the shear occurs by the flux at the membrane surface, but as a result, there is a fall on transmembrane pressure. On the other side, in rotating modules the shear rate is unattached to the feed flow. This way, the shear rate is only about the rotating speed, which induces to an application of these modules in processes with high load of suspended solids, allowing a lasting continuity of permeate quality. The application of these modules is to MF and UF, and can be applied to the biotechnological area, on the treatment of domestic effluents and on the chemical industry (on black liquor treatment) (Serra and Wiesner 2000; Brou et al. 2002; Bhattacharjee and Bhattachary 2006).

6.7 Which Module to Use?

The membranes require a full understanding of various intrinsic characteristics, as, for example, the type of material to be used, in which process the membrane applies and among others, and, clearly, we see importance in choosing the membrane's module. The module choice should consider its application and existence in the commercial field, as well as the needs imposed by the process. If the module presents peculiar characteristics that destine it to specific processes, this module can present a higher added value. Another aspect to be considered is the ease of handling, allowing ease of cleaning and resistance to the gathering of contamination during the process. Besides, membranes that present a different geometry need more specific modules and this sometimes results in higher cost. Finally, the module should adequate easily to the process, presenting flexibility in its scheduling and adequacy to previous or posterior processes to this one.

References

Arndt, F., Ehlen, F., Schütz, S., Anlauf, H., Nirschl, H.: Influence of operating parameters and membrane materials on fouling of ceramic hollow fibre membranes. Sep. Purif. Technol. **171**, 289–296 (2016)

Basile, A., Figoli, A., Khayet, M.: Pervaporation, Vapour Permeation and Membrane Distillation: Principles and Applications. Elsevier Science, Cambridge (2015)

Bhattacharjee, C., Bhattachary, P.K.: Ultrafiltration of black liquor using rotating disk membrane module. Sep. Purif. Technol. **49**, 281–290 (2006)

Bouzerar, R., Paullier, P., Jaffrin, M.Y.: Concentration of mineral suspensions and industrial effluents using a rotating disk dynamic filtration module. Desalination **158**, 79–85 (2003)

Brou, A., Ding, L.H., Jaffrin, M.Y.: Dynamic microfiltration of yeast suspensions using rotating disks equipped with vanes. J. Membr. Sci. **197**, 269–282 (2002)

Busch, J., Cruse, A., Marquardt, W.: Modeling submerged hollow-fiber membrane filtration for wastewater treatment. J. Membr. Sci. **288**(1–2), 94–111 (2007)

Escobar, I., Bruggen, B.V.: Modern Applications in Membrane Science and Technology. Oxford University Press, Oxford (2011)

Govardhan, B., Chandrasekhar, S.S., Sridhar, S.: Purification of surface water using novel hollow fiber membranes prepared from polyetherimide/polyethersulfone blends. J. Environ. Chem. Eng. **5**(1), 1068–1078 (2017)

Ho, W.S.W., Sirkar, K.: Membrane Handbook. Van Nostrand Reinhold, New York (1992)

Hoek, E.M.V., Tarabara, V.V.: Encyclopedia of Membrane Science and Technology. Wiley, Hoboken (2013)

Hubadillah, S.K., Othman, M.H.D., Harun, Z., Ismail, A.F., Iwamoto, Y., Honda, S., Rahman, M. A., Jaafar, J., Gani, P., Sokri, M.N.M.: Effect of fabrication parameters on physical properties of metakaolin-based ceramic hollow fibre membrane (CHFM). Ceram. Int. **42**, 15547–15558 (2016)

Karabelas, A.J., Kostoglou, M., Koutsou, C.P.: Modeling of spiral wound membrane desalination modules and plants: review and research priorities. Desalination **356**, 165–186 (2015)

Li, N.N., Fane, A.G., Ho, W.S.W., Matsuura, T.: Advanced Membrane Technology and Applications. Wiley, New Jersey (2008)

Mulder, M.: Basic Principles of Membrane Technology. Springer Science, The Netherlands (1996)

Pabby, A.K., Rizvi, S.S.H., Sastre, A.M.: Handbook of Membrane Separations: Chemical, Pharmaceutical, Food, and Biotechnological Applications. Taylor & Francis Group, Boca Raton (2009)

Porter, M.C.: Handbook of Industrial Membrane Technology. Noyes Publications, New Jersey (1990)

Purchas, D.B., Sutherland, K.: Handbook of Filter Media. Elsevier, Oxford (2002)

Scott, K.: Handbook of Industrial Membranes. Elsevier, Oxford (1998)

Serra, C.A., Wiesner, M.R.: A comparison of rotating and stationary membrane disk filters using computational fluid dynamics. J. Membr. Sci. **165**, 19–29 (2000)

Shimizu, Y., Okuno, Y., Uryu, K., Ohtsubo, S., Watanabe, A.: Filtration characteristics of hollow fiber microfiltration membranes used in membrane bioreactor for domestic wastewater treatment. Water Res. **30**(10), 2385–2392 (1996)

Wang, L.K., Chen, J.P., Hung, Y.-T., Shammas, N.K.: Handbook of Environmental Engineering: Membrane and Desalination Technologies. Springer, New York (2011)

Wit, P., Daalen, F.S., Benes, N.E.: The mechanical strength of a ceramic porous hollow fiber. J. Membr. Sci. **524**, 721–728 (2017)

Yuliwati, E., Ismail, A.F., Matsuura, T., Kassim, M.A., Abdullah, M.S.: Effect of modified PVDF hollow fiber submerged ultrafiltration membrane for refinery wastewater treatment. Desalination **283**, 214–220 (2011)

FSC
www.fsc.org
MIX
Papier | Fördert gute Waldnutzung
FSC® C083411

Kolibri 360 GmbH
Ferdinand-Jühlke-Straße 7
99095 Erfurt, Deutschland
produktsicherheit@kolibri360.de